Introduction to Machining Science

Introduction to Machining Science

(Second Edition)

G.K. Lal
Department of Mechanical Engineering
Indian Institute of Technology
Kanpur

PUBLISHING FOR ONE WORLD
NEW AGE INTERNATIONAL (P) LIMITED, PUBLISHERS
New Delhi • Bangalore • Chennai • Cochin • Guwahati • Hyderabad
Jalandhar • Kolkata • Lucknow • Mumbai • Ranchi
Visit us at **www.newagepublishers.com**

Published by New Age International (P) Ltd., Publishers
First Edition: 1996
Second Edition: 2007
Reprint: 2015

BRANCHES

- **Bangalore** 37/10, 8th Cross (Near Hanuman Temple), Azad Nagar, Chamarajpet, Bangalore-560 018 Tel.: (080) 26756823, Telefax: 26756820, **E-mail: bangalore@newagepublishers.com**
- **Chennai** 26, Damodaran Street, T. Nagar, Chennai-600 017, Tel.: (044) 24353401 Telefax: 24351463, **E-mail: chennai@newagepublishers.com**
- **Cochin** CC-39/1016, Carrier Station Road, Ernakulam South, Cochin-682 016 Tel.: (0484) 2377303, Telefax: 4051303, **E-mail: cochin@newagepublishers.com**
- **Guwahati** Hemsen Complex, Mohd. Shah Road, Paltan Bazar, Near Starline Hotel, Guwahati-781 008 Tel.: (0361) 2513881, Telefax: 2543669, **E-mail: guwahati@newagepublishers.com**
- **Hyderabad** 105, 1st Floor, Madhiray Kaveri Tower, 3-2-19, Azam Jahi Road, Near Kumar Theater Nimboliadda, Kachiguda, Hyderabad-500 027, Tel.: (040) 24652456, Telefax: 24652457 **E-mail: hyderabad@newagepublishers.com**
- **Kolkata** RDB Chambers (Formerly Lotus Cinema) 106A, 1st Floor, S.N. Banerjee Road, Kolkata-700 014 Tel.: (033) 22273773, Telefax: 22275247, **E-mail: kolkata@newagepublishers.com**
- **Lucknow** 16-A, Jopling Road, Lucknow-226 001, Tel.: (0522) 2209578, 4045297, Telefax: 2204098 **E-mail:lucknow@newagepublishers.com**
- **Mumbai** 142C, Victor House, Ground Floor, N.M. Joshi Marg, Lower Parel, Mumbai-400 013 Tel.: (022) 24927869, Telefax: 24915415, **E-mail: mumbai@newagepublishers.com**
- **New Delhi** 22, Golden House, Daryaganj, New Delhi-110 002, Tel.: (011) 23262368, 23262370 Telefax: 43551305, **E-mail: sales@newagepublishers.com**

ISBN: 978-81-224-2104-0

₹ 299.00

C-15-04-8369

Printed in India at Glorious Printers, Delhi.
Typeset at Aakriti Graphics, Delhi.

PUBLISHING GLOBALLY
NEW AGE INTERNATIONAL (P) LIMITED, PUBLISHERS
7/30 A, Daryaganj, New Delhi-110002
Visit us at **www.newagepublishers.com**

Preface to the Second Edition

In this edition, while retaining the analytical approach to the mechanics of machining, several additions have been made. Two new chapters on 'Mechanics of Oblique Cutting' and 'Vibrations in Machining' have been added. The oblique cutting chapter extends the concept developed for orthogonal cutting to three-dimensional cutting using a wedged-shaped tool. The vibration chapter discusses the types of vibration developed during machining and their causes. Machine tool chatter problem has been analyzed and stability conditions have been derived for minimizing this problem. The chapter on 'Mechanics of Orthogonal Cutting', has also been extended to include two other analyses of orthogonal cutting. In the 'Tool Wear and Tool Life' chapter, a new section on 'Quick Tool-life Testing' has been added. Relevant analytical details have also been included to explain the basic concept involved in quick tool-life testing. Numerical examples, review questions and bibliography have been accordingly extended to include the new topics covered in this edition. It is expected that these additions will benefit the students as well as the practicing engineers in understanding some of the practical phenomena and in predicting the machining performance.

The author would like to thank Dr. A.K. Mallik, Dr. N. Venkata Reddy and Dr. S.K. Choudhury of the Deartment of Mechanical Engineering, I.I.T. Kanpur for their suggestions and assistance during the preparation of this edition. Financial assistance from AICTE during the preparation of this manuscript is gratefully acknowledged.

G.K. Lal

Preface to the First Edition

Machining processes have been in existence for a long time but until the beginning of the present century, improvements in these processes were gradual and came about more or less on a trial-and-error basis. Early 40s marked the beginning of an active interest in the scientific study of machining processes and the aplication of these results have brought in major improvements in tool design, tool materials and machining techniques. These developments have relied heavily on the understanding of fundamentals of metal machining through analyses.

Realizing the importance of metal cutting, many universities have introduced various courses in this area. There is, however, a great deal of diversity in their contents. This book is an attempt to standardize the course material and to emphasize on the fundamental mechanics of the machining process, using an analytical approach. The chapters have been arranged in a logical sequence with contents designed to be covered in a course of about 40 lectures. The presentation is such that it requires no special background and can be easily followed by practicing mechanical engineers. The background covered would be specially useful for tool and product design.

The book covers almost all aspects of machining science that are suitable for senior undergraduate students. Details of machining operations and machine tools have not been included and the readers are assumed to have this basic background. The analyses presented have been further illustrated through numerical examples and a set of review questions is included for evaluating one's understanding of the subject. The illustrations and the line diagrams have been specially prepared to facilitate their reproduction on blackboards and handouts. The bibliography indicates the source materials and recommended research papers for further reading.

I developed interest in the manufacturing area some 30 years ago as a postgraduate student and have been very fortunate in having the association of well-reputed teachers, talented colleagues and enthusiastic students—several of them have contributed directly or indirectly in the preparation of this book. I would like to express my gratitude to Dr. A.P. Verma, a former student and a colleague, for providing enormous assistance and many useful suggestions during the preparation of the manuscript. Thanks to Dr. A.K. Srivastava, for his contributions in the development of the manuscript. Financial assistance for the preparation of the manuscript was provided by Ministry of Human Resource Development, Government of India through its Quality Improvement Programme at I.I.T. Kanpur. This assistance is gratefully acknowledged. I am also grateful to Messers B.L. Arora, C.M. Abraham and J.S. Rawat for their assistance in the preparation of figures and typing of the manuscript.

I would like to dedicate this book to the memory of my father, Dr. K.N. Lal, who always inspired me to write but did not live to see the completion of this book. The book would never have been completed without the encouragement of my wife Dr. (Mrs.) Karuna Lal and daughters Swati, Shruti and Shachi.

G.K. Lal

Contents

Nomenclature

CHAPTER 1

A	Cross-sectional area
A_o	Original cross-sectional area
a, b	Constants
$e, e', e'', e_1, e_2, e_3, e_x, e_y$	Direct strain
e_c	Critical strain
e^e	Elastic strain
e^p	Plastic strain
$\dot{e}$	Strain rate
G_t	Toughness
l, l_1, l_2	Length
l_o	Original length
n	Strain-hardening exponent
P	Load
W_1, W_2	Width
W_p	Plastic work
Z	Critical sub-tangent
γ_{yz}, γ_{zx}	Shear strains
σ	Stress
θ	Temperature

CHAPTER 2

i	Inclination angle
v	Cutting velocity
α	Rake angle
δ	Flank angle

CHAPTER 3

i	Inclination angle
r	Nose radius
x_f, y_f	Longitudinal and transverse feed directions

(x, y, z)	System of axes (subscript a: ASA, o: ORS, m: MRS, n: NRS, au: auxiliary)
α	Rake angle (subscript b: back, e: effective, m: maximum, n: normal, o: orthogonal, r: radial, s: side, v: velocity rake angles)
γ	Cutting-edge angle (subscript a: auxiliary, e: end, p: principal, s: side cutting-edge angles)
δ	Flank angle (subscript a: auxiliary, an: auxiliary normal, e: end, p: principal, pn: principal normal, s: side flank angles)
θ_m	Oblique plane angle
π	Reference plane (subscript an: auxiliary normal, c: cutting, m: oblique, r: oblique, r: base, x: longitudinal, y: transverse planes)
ψ	Helix angle

CHAPTER 4

A, A_r	Apparent and real areas of contact
A_s	Area of shear plane
b, b'	Width of workpiece
b_c	Width of chip
F	Friction force
F_{cp}, F_{cq}	Clearance face forces in the tangential and normal directions
F_p, F_q	Tangential and normal forces
F_S, F_N	Shear force and normal force on the shear plane
h	Chip-tool contact length
k	Yield shear stress
l	Length of cut
l_c	Chip length
N	Normal force on rake face
p_A, p_B	Hydrostatic stress
R, R'	Resultant force
t, t_c	Undeformed and deformed chip-thicknesses
V, V_c, V_s	Cutting, chip and shear velocities
W_c	Weight of chip
Z, Z'	Moment arm
α	Rake angle
γ	Shear strain
$\dot{\gamma}$	Shear strain rate
δ	Flank angle
λ	Friction angle
μ	Coefficient of friction
ρ	Density
σ	Normal stress on rake face
$\sigma_1, \sigma_2, \sigma_3,$	Principal stresses
τ	Shear stress on rake face
ψ	Inclination of slip line with reference axis
ϕ	Shear angle

CHAPTER 5

A_s	Area of shear plane
b, b_c	Width of workpiece and chip
F, F'	Friction forces
$F_P, F_Q, F_R, F'_p, F'_Q, F'_R$	Force components in P, Q and R directions
F_S, F_N	Shear and normal forces on shear plane
i	Inclination angle
k	Yield shear stress
l	Length of cut
l_c	Chip length
R, R', R''	Resultant forces
R_T, R_N	Rake face forces
r_t, t_b	Chip-thickness and chip-width ratios
S_T, S_N	Shear plane forces
t, t_c	Undeformed and deformed chip-thicknesses
V_c, V_w, V_s	Chip, cutting and shear velocities
W_T, W_N	Flank face forces
$\alpha_e, \alpha_n, \alpha_v$	Effective, normal and velocity rake angles
$\gamma, \dot{\gamma}$	Shear strain and shear strain rate
λ, λ_n	Friction and normal friction angles
τ, σ	Shear and normal stresses
η_c, η_s	Chip-flow and shear-flow angles
ϕ_e, ϕ_n	Effective and normal shear angles

CHAPTER 6

A	Area of cross-section
b	Width
e, e_1, e_2, e_3, e_4	Strain
E	Modulus of elasticity
F, F_p, F_q, F_r	Force (Subscripts p, q and r refer to p, q and r directions)
G	Gauge factor
h	Depth
i_1, i_2, i_3, i_4, i_g	Current
K_s	Spring constant
L, l	Length
M	Moment
n	Number of active arms
r	Radius of ring
$R, R_1, R_2, R_3 R_4$	Resistance of strain gauges

R_x, R_y, R_z	Recorder output in *x*, *y* and *z* directions
t	Thickness of ring
V_e	Impressed voltage
v'	Output voltage
W	Wattage
ν	Poisson's ratio

CHAPTER 7

A	Cross-sectional area
a	Constant
b	Width of workpiece
c	Specific heat
E_m	Average emf at the contacting surface
E_n	Junction emf
F, F_p, F_q, F_s	Friction, tangential, thrust and shear forces
K	Thermal conductivity
Q	Heat flow by conduction (subscript *x* and *lf* refer to *x* and *y* directions)
Q'	Heat transferred by transportation (subscript *x* and *y* refer to *x* and *y* directions)
Q_a	Heat absorbed
Q_r	Heat generated
q_r	Rate of heat generation
q_f	Rate of heat generation due to rake face friction
q_s	Rate of heat generation on the shear plane
R_n	Junction resistance
R_T	Thermal number
t, t_c	Undeformed and deformed chip-thicknesses
V	Velocity (subscript *x* and *y* refer to *x* and *y* directions)
V_c, V_s	Chip and shear velocities
W_s	Shear energy
α	Rake angle
α_d	Thermal diffusivity
β	Fraction of heat going into chip
θ	Instantaneous temperatures
$\bar{\theta}$	Average temperature
θ_c	Temperature across the chip
θ_f	Temperature increment in the chip due to friction
$\bar{\theta}_f$	Average chip-tool interface temperature
θ_o	Ambient temperature
θ_s	Shear plane temperature
$\bar{\theta}_s$	Average shear plane temperature

ρ	Density
τ	Time
ϕ	Shear angle

CHAPTER 8

C, C', C_1, C_2	Tool life constants
d	Depth of cut
d_c	Depth of crater
D_1, D_2	Diameter
f	Feed rate, Feed per revolution
H	Hardness
L	Length of workpiece
l_w	Wear land size
l_w*	Critical wear land
N_1, N_2	Speed (rpm)
R, R_1, R_2	Radius
T	Tool life
T_1	Time
V	Cutting speed
V_{eq}	Equivalent speed
V_T	Cutting speed for a given tool life
m	Constant
n, n_1, n_2	Tool life exponents
X, X_1, X_2	Fraction of tool life
α	Rake angle
γ	Side cutting-edge angle
δ	Flank angle
θ	Tool temperature
σ	Yield stress

CHAPTER 9

$\alpha_1, \alpha_2, B_p, B_s, B_w', C$	Constants
C_B	Machining cost per batch
C_C	Cutting cost per batch
C_T	Tool cost
C_N	Non-productive cost
C_g	Tool regrinding cost
C_i	Initial cost of tool
C_p	Machining cost per piece
c_D	Depreciation cost rate
c_L	Labour cost rate
c_O	Overhead cost rate

c_R	Total cost rate
D	Diameter of workpiece
d	Depth of cut
F_p	Cutting force
f	Feed rate
h_m	Maximum permissible surface roughness
I_p	Income per piece
K_1, K_2,	Constants
L	Length of workpiece
m_1, m_2, n	Constants
N_B	Number of components in a batch
N_g	Number of components produced between each tool regrind
P_R	Profit rate
P_w	Power required
R_P	Production rate
r_g	Number of regrinds possible on a tool
T	Tool life
T_a	Tool advance and withdrawal time
T_B	Time for machining a batch of components
T_c	Cutting time per component
T_i	Idle time
T_l	Loading and unloading time per piece
T_N	Non-productive time
T_p	Machining time per piece
T_r	Tool removing and replacing time
T_s	Initial set-up time

CHAPTER 10

A_c	Cross-sectional area of chip
A_p, A_q, A_r	Constants
B	Width of workpiece
b	Width of chip
C_1, C_2	Constants
D	Diameter of cutter
$\overline{D}$	Mean diameter fo workpiece in turning
D_w	Workpiece diameter
d	Depth of cut
F_P, F_Q	Total tangential and thrust forces during drilling
$F_p, F_q, F_r, F_p', F_q', F_r'$	Force components (subscripts p, q and r refer to p, q and r directions)
F_{p_1}, F_{q_1}	Tangential and thrust forces due to friction at the chisel head
F_{p_2}, F_{q_2}	Tangential and thrust forces due to extrusion at the chisel head
f	Feed rate
h	Peak-to-valley height

i	Inclination angle
l	Chip length
l_s	Sample length
N	Cutter rpm
N_w	Workpiece rpm
P_w	Power
R	Resultant force
r	Nose radius
r_a	Distance from drill axis
s	Feed per tooth
t	Undeformed chip-thickness
$\bar{t}$	Mean undeformed chip-thickness
t_m	Maximum undeformed chip-thickness
u_s	Specific energy
V	Cutting speed
$\bar{V}$	Mean cutting speed
v	Feed velocity
$x_1, x_2, x_p, x_q, x_r,$ $y_1, y_2 y_p, y_q, y_r$	Constants
Z_m	Material removal rate
α	Rake angle (subscripts n and o refer to normal and orthogonal rake angles)
γ	Cutting edge angle (subscript e: end, p: principal and s: side cutting-edge angles)
τ	Time

CHAPTER 11

A_c	Cross-sectional area of chip
B	Width of grinding wheel in contact with workpiece
b	Width of chip
C	Number of cutting points per unit wheel surface area
D	Diameter of wheel
D_w	Diameter of workpiece
d	Infeed
F_g	Force on an individual grain
F_p, F_q, F_r	Force components in p, q ad r directions
G	Grinding ratio
g	Grain diameter
h	Mean peak-to-valley height
K	Thermal conductivity
K_g	Constant
l	Chip length (subscript e, i and s refer to external cylindrical, internal and horizontal surface grinding operations)
m	Specific energy exponent

q	Heat source
$\bar{q}$	Heat source in grinding
r	Ratio of chip width and mean chip-thickness
s_f	Distance moved by workpiece between passage of successive grains
s_n	Sieve number
t	Maximum undeformed chip-thickness
u	Specific energy
V	Wheel speed
v	Work speed
W_M	Volume of material removed
W_W	Volume of wheel wear
α_d	Thermal diffusivity
θ	Temperature
$\bar{\theta}$	Mean temperature
θ_m	Maximum temperature
τ	Time

CHAPTER 12

A, A_1, A_2	Constants
$F, F(t)$	Force
k	Spring stiffness
k_1, k_2	Damping
m	Mass
t, t'	Time
x	Acceleration
$\alpha, \alpha_1, \alpha_2, \beta, \beta_1, \beta_2$	Angles
ψ	Constant
w, w_0	Forcing frequency
ξ	Damping factor
γ	Coupling coefficient

CHAPTER 1

Deformation of Metals

1.1 INTRODUCTION

The cutting process has been used by mankind since ancient days. The earliest cutting tool was perhaps a blunt chisel but with experience the importance of tool angles was realised. The basic features involved in cutting with a chisel or a wedge-shaped tool are common in all machining operations. Earlier, it was believed that when metal is cut, the material merely splits-off in front of the tool somewhat similar to splitting of a fibrous material like wood while cutting with an axe. Subsequently it was realised that cutting of metal involved a deformation process although rupture was considered to be the dominant feature. Further experiments indicated that the deformation was principally one of shear and that the type of chip formed varied with work material and cutting conditions. Extensive research on metal cutting during the past 50 years has revealed the complexity of the process. The process essentially involves plastic deformation and fracture under high strain rate and high temperature conditions not normally encountered in available material testing methods. It is, therefore, desirable to outline the basic features of metal deformation and the mechanism involved before considering the cutting operations.

1.2 PLASTIC DEFORMATION

When materials are subjected to external loads, they are distorted or deformed. This deformation may be elastic, plastic or fracture. When the material returns to its original configuration on removal of external loads, the deformation is elastic in nature but when it does not return to its original configuration and remains as a continuous mass, plastic deformation is said to have occurred. During fracture, a part of the original body of material is completely separated from the rest.

1.2.1 Tensile Test

A simple tensile test on a mild steel bar is perhaps the most familiar example of elastic and plastic deformation and fracture. In a tensile test, the load on the specimen is gradually increased and the corresponding increase in length is recorded. It is found that initially the increase in length is proportional to the applied load, and when the applied load is removed the bar returns to its original length. Thus, in this region the metal exhibits elastic behaviour. This is the linear range *OA* of the curve in Fig. 1.1. On further straining, the relationship between load and extension no longer remains linear but the material is still elastic. The maximum load that can be applied without causing permanent deformation defines the elastic limit. Usually there is a little difference between the proportional limit *A*, and the elastic limit *B*. Beyond this limiting load the metal no longer behaves elastically and removal of the load leaves the bar with a permanent extension. The point at which the elastic behaviour of the material ceases and yielding starts is called the *yield point*. Yielding, especially in mild steel, is accompanied by a sudden drop in load and increase in length at approximately constant load. This is the region *CD* of the curve. The stress at *B* is usually called the *upper yield stress* and that at *C* the *lower yield stress*. This type of yielding, showing upper and lower yield stresses, is characteristic of mild steel and is caused by the presence of small amount of foreign atoms in the material, usually carbon or nitrogen. Many of these foreign atoms migrate to regions of higher energy such as dislocation and anchor the dislocation movement giving upper yield stress during yielding. Most metals and alloys do not show this type of pronounced initial yielding and the change from elastic to elastic-plastic deformation is gradual.

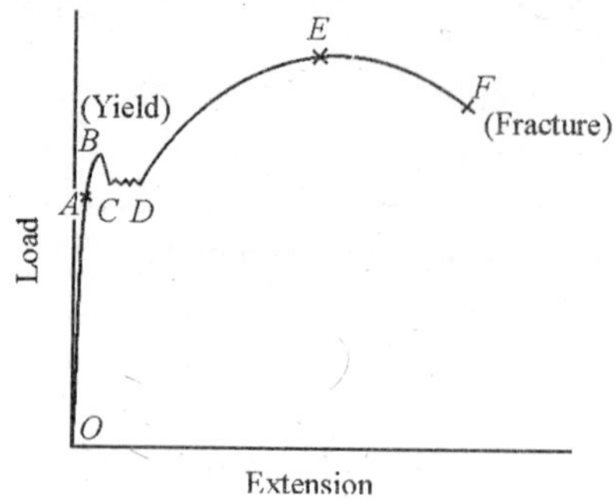

Fig. 1.1 Load-extension curve for mild steel

Following the transition from elastic to plastic behaviour, the load required for further deformation increases but is no longer related by a constant of proportionality as in the elastic range where the constant is defined as the modulus of elasticity. This effect of the material being able to withstand the increased load despite the uniform reduction in cross-sectional area is called strain-hardening or work-hardening. Two opposing factors operate to determine the load required for a given extension. Strain-hardening leads to an increase of load while the corresponding reduction in cross-sectional area leads to a decrease of load. At some point *E*, when the extension becomes large, the strain-hardening is unable to cope up with the decrease in cross-sectional area and a maximum occurs in the load followed by localized deformation in the specimen and is indicative of an instability condition. This leads to a decrease in the cross-sectional area and increase in the local stress so that further elongation occurs in the thinned portion only and a 'neck' is developed. From the point of instability at *E* to fracture at *F*, straining takes place under a complicated and continuously changing triaxial tensile stress system.

1.2.2 True Stress and True Strain

Nominal stress is defined as the load divided by the original cross-sectional area of the bar and the engineering strain as the extension per unit original length. The load extension diagram for a bar with unit cross-sectional area and unit gauge length is therefore the nominal stress-strain curve for the material.

The nominal stress defined in terms of the original cross-sectional area is not really a stress since the cross-sectional area A at the instant of load measurement is less than the original cross-sectional area A_0. In the elastic region $A \approx A_0$, but in the plastic region where the metal working operations are carried out this is far from true. If we neglect the very small change in volume during a tensile test and assume that the material is incompressible, then

$$Al = A_0 l_0 \qquad \ldots(1.1)$$

where A is the current cross-sectional area and l is the current gauge length. Suffix 0 refers to the original conditions. If P is the current load, the true or natural stress σ is given by

$$\sigma = \frac{P}{A} \qquad ...(1.2)$$

The true or natural strain e is defined in terms of the strain increment, that is

$$e = \int_{l_0}^{l} \frac{dl}{l} = \ln\left(\frac{l}{l_0}\right) \qquad ...(1.3)$$

There are two main advantages in using this logarithmic definition of strain:

(*i*) Logarithmic strains are additive in sequential processes, which is not true of linear strains. Suppose a bar of length l_0 is strained during a tensile test to a length l_1. The true strain $e' = \ln\left(\frac{l_1}{l_0}\right)$. If the bar is further extended from length l_1 to length l_2, the further strain will be $e'' = \ln\left(\frac{l_2}{l_1}\right)$. The total strain e is evaluated as

$$e = e' + e''$$

$$= \ln\left(\frac{l_1}{l_0}\right) + \ln\left(\frac{l_2}{l_1}\right)$$

$$= \ln\left(\frac{l_2}{l_0}\right) \qquad ...(1.4)$$

(*ii*) In metal working operations, the material is subjected to large plastic strains and under such conditions the material may be considered to be incompressible. This volume constancy condition in terms of true or logarithmic strains can be simply written as

$$e_1 + e_2 + e_3 = 0 \qquad ...(1.5)$$

where e_1, e_2, e_3 are logarithmic or true strains in three mutually perpendicular directions.

1.2.3 Stress-Strain Curves

The load extension relationship is usually expressed as an engineering stress-engineering strain curve and there is a maximum value of stress, usually called the ultimate tensile strength. The material in fact continues to strain-harden upto the point of fracture and the stress required for further deformation should also increase. This is the case when the true stress and true strain curve is plotted as shown in Fig. 1.2.

Several attempts have been made to suggest empirical equations for this curve. The expression most commonly used is of the form

$$\sigma = a\,(b + e)^n \qquad ...(1.6)$$

where a, b and n are constants for a given material and can be evaluated experimentally.

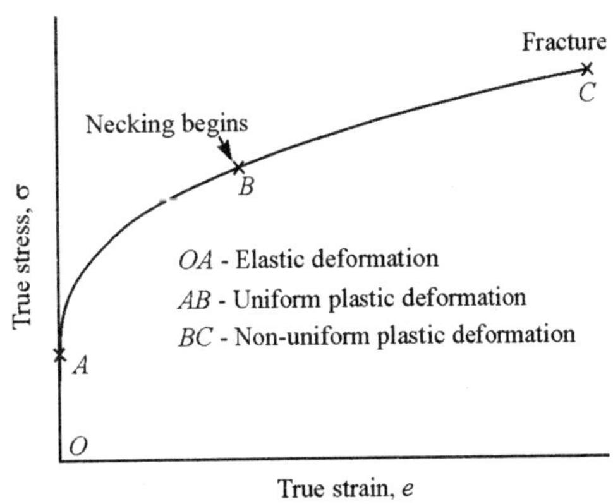

Fig. 1.2 True stress–true strain curve

In order to obtain a solution to a deformation problem, it is often necessary to idealize this stress-strain relation. Such idealization will simplify the solution but its field of application will be restricted. The total strain e is defined as the sum of the elastic and the plastic strain components. Thus,

$$e = e^e + e^p \qquad \ldots(1.7)$$

where e^e is the elastic strain and e^p is the plastic strain. Elastic strains in deformation processes are usually very small (about 0.1 percent), whereas plastic strains are seldom less than 10 percent. Therefore, elastic strains can be neglected in computing the total strain and $e \approx e^p$. Often, for simplification the effect of work-hardening is also ignored. Thus, the stress-strain curves can be idealized and we may describe in turn

(*a*) a rigid, perfectly plastic material
(*b*) an elastic, perfectly plastic material
(*c*) a rigid, linear strain-hardening material and
(*d*) an elastic, linear strain-hardening material

These are shown in Fig. 1.3.

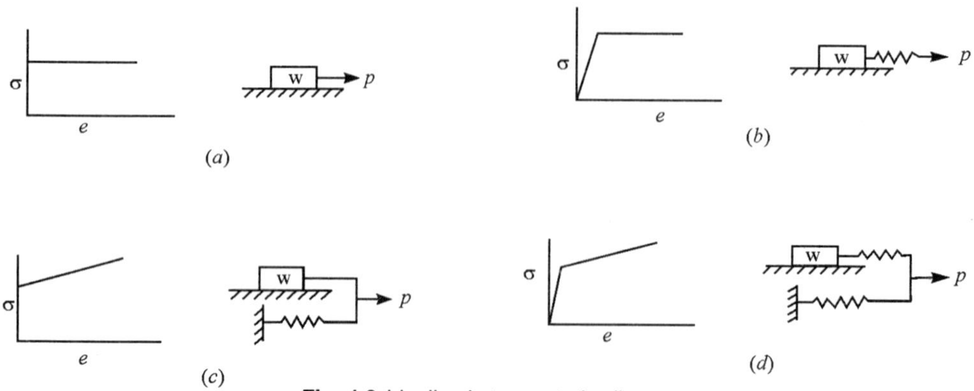

Fig. 1.3 Idealized stress-strain diagrams
(a) Rigid, perfectly plastic. (*b*) Elastic, perfectly plastic.
(*c*) Rigid, linear strain-hardening. (*d*) Elastic, linear strain-hardening.

A simple mechanical analogy exists between the behaviour of these idealized materials and the motion of a spring loaded block. In Fig. 1.3(*a*), the block can move only when the applied force exceeds the friction force, thus representing a rigid, perfectly plastic material. Once the movement begins, it will continue to slide under a constant force. In Fig. 1.3(*b*), the deflection of the spring is proportional to the applied force and represents the elastic region of the curve. Once the force exceeds a certain amount to overcome friction, the block will continue to slide under constant force. This model will represent the mechanical behaviour of an elastic, perfectly plastic material. A rigid, linear strain-hardening material, and an elastic, linear strain-hardening material, can similarly be represented by spring and block arrangements shown in Figs. 1.3(*c*) and 1.3(*d*).

1.2.4 Some Deviations in Stress-Strain Curves

The stress-strain curve of a material depends among other things, on the stress system, rate of straining and temperature. For example, if in a simple tensile test, the applied load is removed after the material has exceeded the yield stress and loaded again, a hysteresis loop is formed as shown in Fig. 1.4. On unloading at a point *A* beyond the yield stress the material follows a curve *ABC*, almost parallel to the elastic loading path. On reloading from point *C*, it follows the path *CDY** giving a *hysteresis loop*.

Yielding now occurs at Y^* giving a value larger than the original yield stress value at Y, and the curve Y^*E is almost in continuation of the earlier path YA.

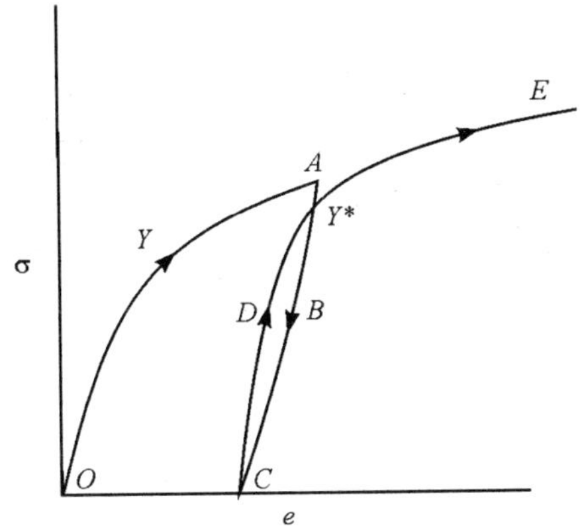

Fig. 1.4 Hysteresis effect

Similarly, if a metal is plastically deformed in tension upto a point A (Fig. 1.5) and unloaded to point B and then subjected to a compressive load, yielding occurs at Y_c giving a lower yield stress value than that at Y. This is because of the residual stresses developed in the metal due to loading and unloading. This is the *Bauschinger effect* and is present in deformation processes involving reversal of stress.

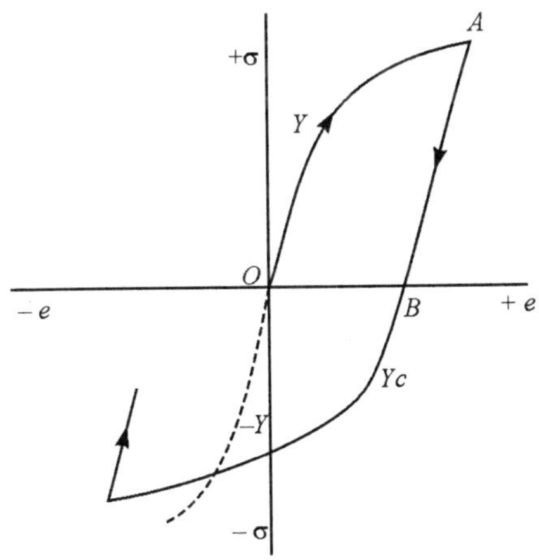

Fig. 1.5 Bauschinger effect

When metal is plastically deformed, the grains which are initially oriented at random (isotropic) get a preferred orientation and become anisotropic. This causes the yield stress to depend on the direction in which they are measured. For example, in a cold rolled strip of metal, the tensile yield stress transverse to the direction of rolling may be greater than the tensile yield stress in the rolling direction.

1.2.5 Temperature and Strain Rate Effects

The stress-strain curve of a material is sensitive to both temperature and rate of strain. In general, increase in temperature (θ) reduces the level of yield stress (Fig. 1.6) and also tends to reduce strain-hardening which results in flattening of the curve. Thus, fracture occurs with increased ductility. On the other hand, time-rate of strain $(\dot{e})$ tends to increase the level of yield stress.

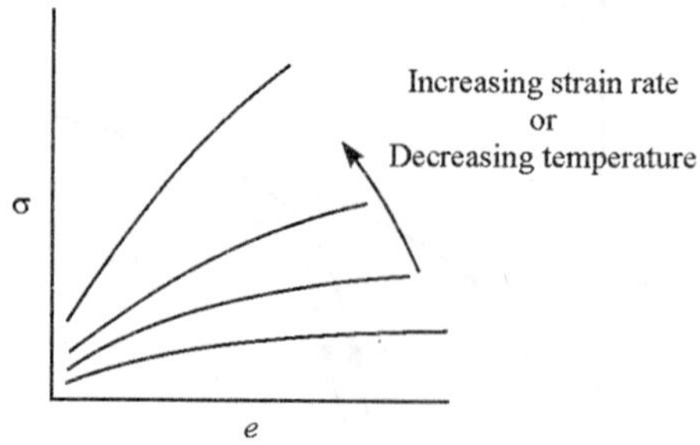

Fig. 1.6 Strain rate and temperature effects

Various attempts have been made to relate yield or flow stress for a given strain and strain-rate to temperature. A simple relationship suggested is of the form

$$\sigma(e,\dot{e}) = \text{Constant}/\theta \qquad \text{...(1.8)}$$

Similarly, the most general relationship between flow stress for a given strain and temperature is a power expression of the form

$$\sigma(e,\theta) = \text{Constant } (\dot{e})^m \qquad \text{...(1.9)}$$

where m is known as the strain-rate sensitivity and is generally dependent on temperature. The exponent m can be evaluated by deforming the material at two different strain rates $\dot{e}_1$ and $\dot{e}_2$ at a particular temperature (θ) and measuring the corresponding flow stress σ_1 and σ_2 at some predetermined strain. Then,

$$m = \frac{\ln(\sigma_2/\sigma_1)}{\ln(\dot{e}_2/\dot{e}_1)} \qquad \text{...(1.10)}$$

1.3 DUCTILITY AND TOUGHNESS

The term ductility is commonly used to describe the ability of a metal to undergo plastic deformation before fracture. Similarly, toughness is the ability of a metal to absorb energy in plastic deformation upto the point of fracture. Toughness is therefore a function of both strength and ductility. The limit of useful deformation is, however, a function of the total strain which can be imposed on a metal before necking occurs. In view of this, a suitable measure of useful ductility is the total strain at the onset of necking. Similarly, toughness can be defined in terms of the area under the stress-strain curve upto the onset of necking.

At the point of instability the rate of increase of load due to strain-hardening is just balanced by the rate of decrease due to diminishing area of cross-section. Consider again the tensile test specimen of current cross-sectional area A subjected to a load P. Differentiating the true stress given by

$$\sigma = \frac{P}{A}$$

with respect to the current length l, we get

$$\frac{dP}{dl} = \sigma\frac{dA}{dl} + A\frac{d\sigma}{dl} \qquad \text{...(1.11)}$$

The incompressibility condition gives

$$Al = A_0 l_0 = \text{Constant} \qquad \text{...(1.12)}$$

or

$$A\,dl + l\,dA = 0 \qquad \text{...(1.13)}$$

or

$$\frac{dA}{dl} = -\frac{A}{l} \qquad \text{...(1.14)}$$

Using the above equations, we get

$$\frac{dP}{dl} = -\sigma\frac{A}{l} + A\frac{d\sigma}{dl} \qquad \text{...(1.15)}$$

The first term on the right hand side represents the rate of decrease of load due to changes in the cross-sectional area, while the second term represents the rate of increase in load due to strain-hardening. At the point of instability, the load is maximum, *i.e.*, $\frac{dP}{dl} = 0$. Thus,

$$\frac{d\sigma}{\sigma} = \frac{dl}{l} = de \qquad \text{...(1.16)}$$

or

$$\frac{1}{\sigma}\frac{d\sigma}{de} = \frac{1}{Z} = 1, \qquad \text{...(1.17)}$$

where Z is the critical sub-tangent.

If we now draw a tangent at the point of instability E on the true stress-true strain curve (Fig. 1.7), the length DF will be of value unity. Hence, for a given stress-strain curve, the critical strain e_c at

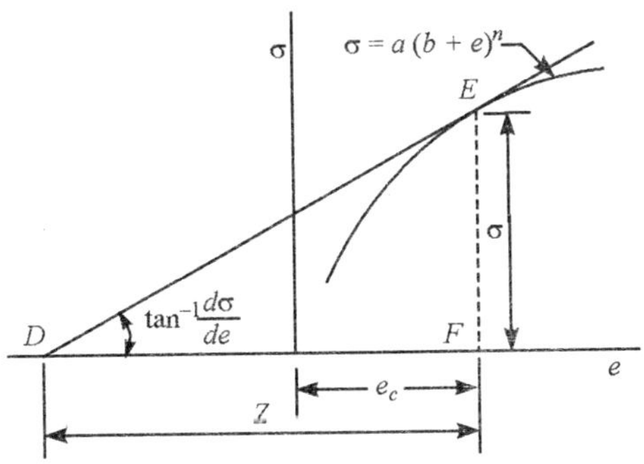

Fig. 1.7 Instability strain

instability can be evaluated. This critical strain is one of the required measure of the inherent ductility of a material. Thus, a material can be classified as ductile or brittle on the basis of their stress-strain characteristics. A ductile material will undergo plastic deformation before fracture (high e_c), while a perfectly brittle material is likely to stay elastic until the fracture point is reached (low e_c).

Assuming a stress-strain curve of the form given by equation (1.6), the critical strain can be evaluated by differentiating this equation and substituting in equation (1.17). This gives,

$$e_c = nZ - b \qquad \text{...(1.18)}$$

The useful ductility is, therefore, a function of both the material properties (b,n) and the critical sub-tangent Z.

The total plastic work done (W_p) upto the point of the instability (onset of necking) is

$$W_p = \int_0^{e_c} \sigma de \quad \ldots(1.19)$$

Using equations (1.6) and (1.18), the useful toughness G_t can be evaluated as

$$G_t = W_p = \frac{a}{n+1}[(nZ)^{n+1} - (b)^{n+1}] \quad \ldots(1.20)$$

When $b = 0$, that is the stress-strain curve passes through the origin, equation (1.20) reduces to

$$G_t = \frac{a}{n+1}(nZ)^{n+1} \quad \ldots(1.21)$$

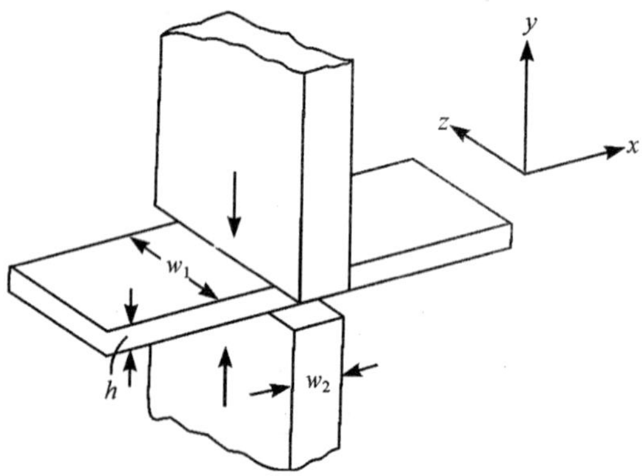

Fig. 1.8 Plane-strain compression

1.4 PLANE STRAIN DEFORMATION

Most metal working problems are analysed in terms of plane strain deformation where the plastic flow occurs entirely in one plane with no deformation in the direction perpendicular to the plane. By way of example, let us consider the compression of a wide strip between two flat-faced dies (Fig. 1.8). The dies overlap the strip in its width direction (W_1) and are narrow in breadth (W_2). When $W_1 > 10\ W_2$ (approximately), lateral strains are negligible and plane strain deformation is obtained. Thus, in plane strain deformation,

(*a*) the deformation is everywhere parallel to a given plane (x,y plane in Fig. 1.8) and

(*b*) the deformation is independent of z.

That is, the plastic strain e_z and shear strain γ_{yz} and γ_{zx} are zero and the consideration of volume constancy gives $e_x = -e_y$.

1.5 MECHANISM OF PLASTIC DEFORMATION

In order to understand the behaviour of metals subjected to large deformation, we must examine the behaviour on the atomic scale. For plastic deformation to take place, it is necessary to have large-scale slipping (more than half atomic spacing) of adjacent atomic planes. During slipping, the structure of the crystal lattice is maintained, but the relative positions of atoms are changed (Fig. 1.9). Every lattice structure has definite slip planes and directions along which slipping can occur. Slipping, however, occurs more readily along planes where the atoms are most closely packed. Figure 1.10 shows the most closely packed atom planes in face centred cubic (*fcc*), hexagonal close packed (*hcp*) and body centred cubic (*bcc*) structures. The strength of such single crystals is determined in terms of shear stress at which slipping occurs.

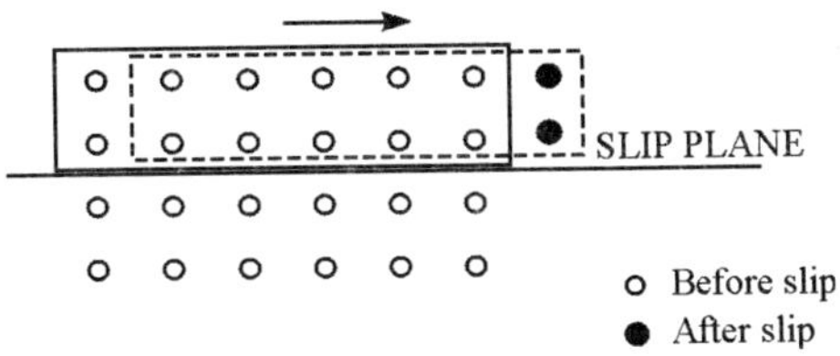

Fig. 1.9 Deformation by slip

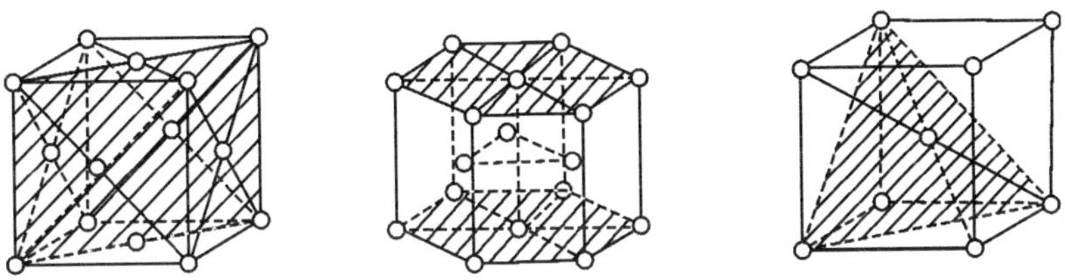

(*a*) (*b*) (*c*)

Fig. 1.10 The most closely packed atom planes in (*a*) face centred cubic, (*b*) hexagonal close packed and (*c*) body centred cubic cell structures

Metals can also be deformed by the process of *twinning*. In this process, the rotation of part of the crystal relative to the other part occurs in such a way that the deformed portion becomes a mirror image of the undeformed portion as shown in Fig. 1.11. Twinning is an abrupt process and requires higher stress for its initiation than does deformation by glide.

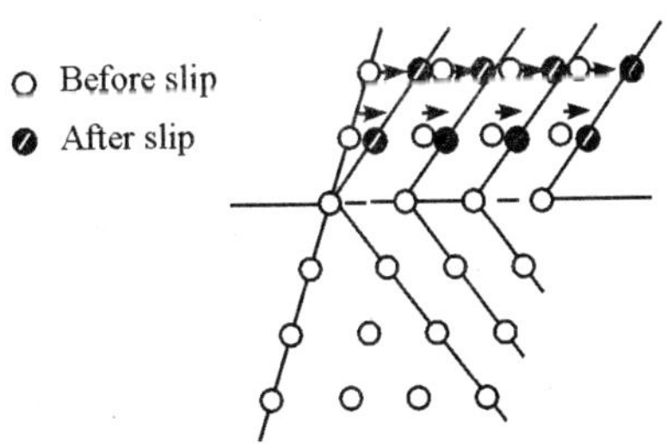

Fig. 1.11 Deformation by twinning

The simplest mechanism by which slipping could occur is through rigid body translation of the material on one side of the slip line relative to that on the other. Figure 1.12(*a*) shows an arrangement where atoms are in equilibrium although there is one extra row of atoms below the slip plane. On application of shear stress, the dislocation centre moves (Fig. 1.12*b*) and is finally out of the field of view (Fig. 1.12*c*). The atoms are now perfectly aligned but the upper row in effect has slipped one atomic spacing relative to the lower row. This type of dislocation is called *edge dislocation* and in this case the dislocation line moves in the direction of slip. A three dimensional view is shown in Fig. 1.13.

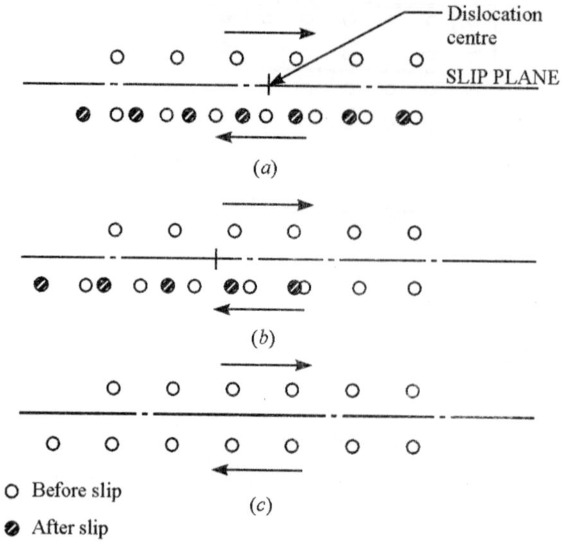

Fig. 1.12 Dislocation motion

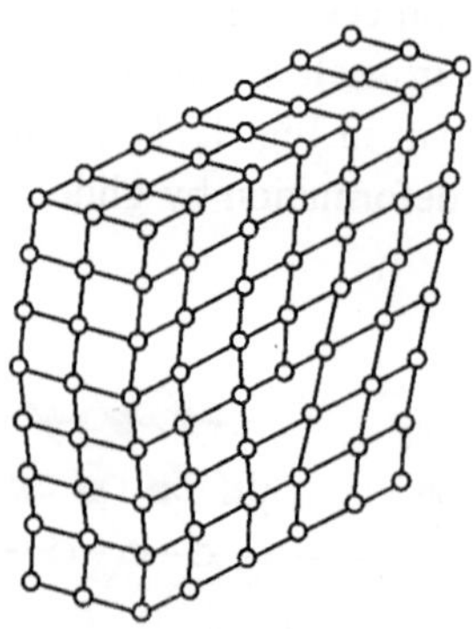

Fig. 1.13 Edge dislocation

When dislocation line moves at right angles to the direction of slip, it is termed as *screw dislocation.* Figure 1.14 illustrates this.

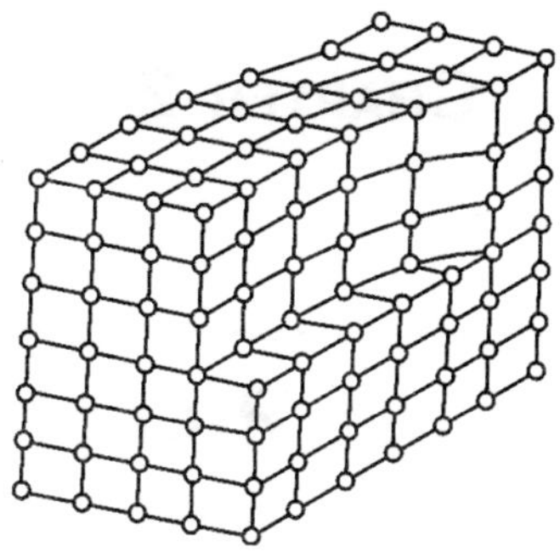

Fig. 1.14 Screw dislocation

The concept of dislocation is able to explain why metals deform at such relatively low stresses in comparison with the stresses required to disturb a perfect lattice.

In practice, slipping, twinning and grain boundary movement may occur simultaneously in several crystals of a polycrystalline metal and in three dimensions. The grain boundary region of polycrystalline materials contains a series of edge dislocations. However, during large deformation it is possible for a dislocation in one crystal to move into a grain boundary and provide a distortion which causes a new dislocation to be formed in the neighbouring grain. In crossing an individual grain, a dislocation encounters resistance due to various irregularities such as foreign atoms, precipitated particles and other dislocations of dissimilar nature present in the crystal at the slip plane. Thus, an increased load is required to cause further movement of the dislocation during slip. This phenomenon is called strain-hardening or the material is said to have work-hardened. Finally, the stress required to cause further glide becomes so high that it attains the cohesive strength of the lattice and fracture occurs.

In reality, there are several possible slip systems even in single crystals because of the defects or imperfections and in engineering metals which are polycrystalline, the slip system may be exceedingly complex. These may include, point defects (vacant atomic site or an interstitial atom), stacking faults (interruption of regular sequence of atoms) and dislocations.

CHAPTER 2

Machining Processes

2.1 INTRODUCTION

The object of any manufacturing process is to produce a piece of specified shape and material properties. Therefore, a process, in general, will produce changes in configuration and physical properties of the raw material. These can be achieved by carrying out one of the following types of operations:

1. Constant mass operations (casting, rolling, extrusion, wire drawing, forging, etc.)
2. Material addition operations (bolting, rivetting, keying, welding, etc.)
3. Material removal operations (machining, grinding, etc.)

2.2 MACHINING PROCESSES

Machining is one of the processes of manufacturing in which the specified shape is imparted to the workpiece by removing surplus material. Conventionally this surplus material from the workpiece is removed in the form of chips by interacting the workpiece with an appropriate tool. This mechanical generation of chips can be carried out by single- or multi-point tools, or by abrasive operations. These are summarized below:

Machining processes

Single-point tool operations	Multi-point tool operations	Abrasive operations
1. Turning	1. Milling	1. Grinding
2. Boring	2. Drilling	2. Lapping
3. Shaping	3. Tapping	3. Honing
4. Planing	4. Reaming	4. Super-finishing
	5. Hobbing	
	6. Broaching	
	7. Sawing	

The process of chip formation in metal-cutting is effected by relative motion between the tool and the workpiece achieved with the aid of a device called *machine tool*. This relative motion can be obtained by a combination of rotatory and translatory movements of either the tool or the workpiece or both. The kind of surface that is produced by the operation depends on the shape of the tool and the path it traverses through the material. When the workpiece is rotated about an axis and the tool is traversed in a definite path relative to the axis, a surface of revolution is generated. When the tool path is parallel to the axis, the surface generated is a cylinder as in straight turning or boring operations (Figs. 2.1 and 2.2). Similarly, planes may be generated by a series of straight cuts without rotating the workpiece as in shaping and planing operations (Fig. 2.3). In shaping, the tool is reciprocated and the workpiece is moved crosswise at the end of each stroke. Planing is done by reciprocating the workpiece and crosswise movement is provided to the tool.

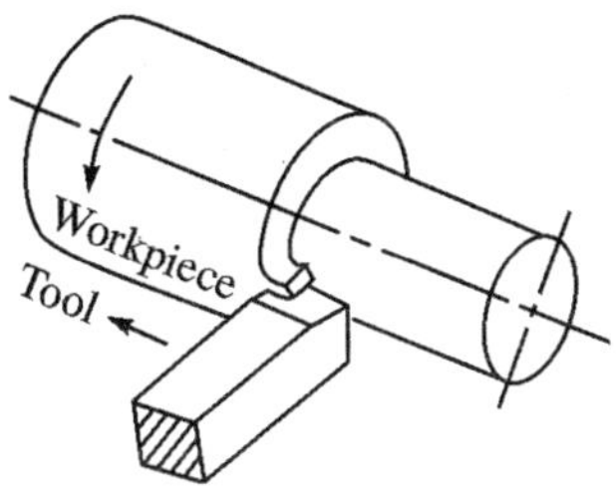

Fig. 2.1 Straight turning

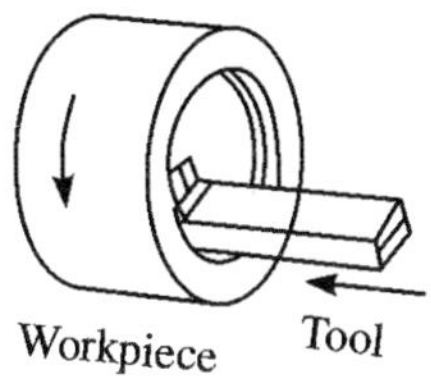

Fig. 2.2 Straight boring

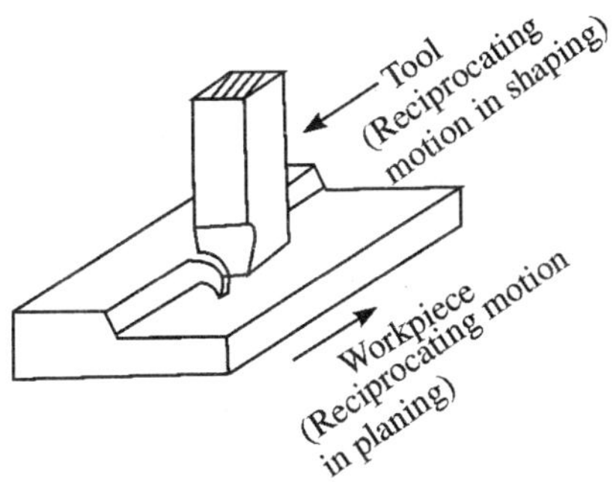

Fig. 2.3 Shaping and planing

Surfaces may be machined by tools having a number of cutting edges that can cut successively through the workpiece material. In plane milling, the cutter revolves and moves over the workpiece as shown in Fig. 2.4. The axis of the cutter is parallel to the surface generated. Similarly in drilling, the drill may turn and be fed into the workpiece or the workpiece may revolve while the drill is fed into it (Fig. 2.5).

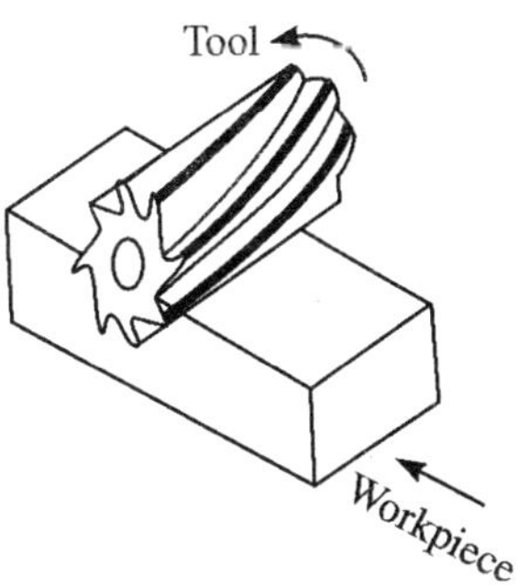

Fig. 2.4 Plain milling

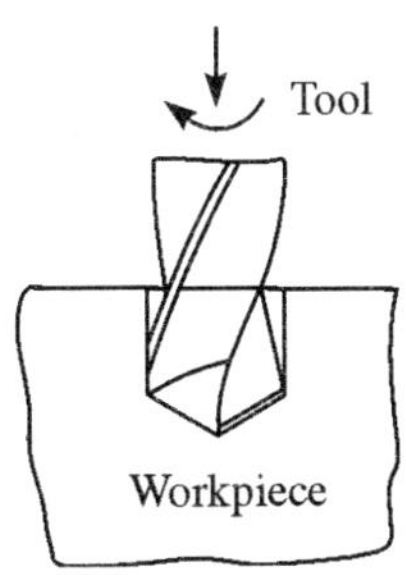

Fig. 2.5 Drilling

The machine tools, in general, provide two kinds of relative motions; the primary motion is responsible for the cutting action and absorbs most of the power required to perform the machining action. The secondary motion or the feed motion may proceed in steps or continuously and absorbs only a fraction of the total power required for machining. When the secondary motion is added to the primary motion, machined surfaces of desired geometric characteristics are produced.

Consider a situation where both the cutting motion as well as the feed motion (provided at the end of each stroke) are rectilinear but perpendicular to each other. Here the machined surface produced is a plane. The line generated by the primary motion (cutting motion) is called the *generatrix*, while the line representing the secondary motion (feed motion) is called the *directrix* (Fig. 2.6*a*).

Depending upon the shapes of the generatrix and the directrix and their relative orientations, various geometries can be produced on the workpiece. Consider another case when the generatrix is a circle and the directrix is a line perpendicular to the plane of the generatrix. It is clear that in this situation the surface produced will be a cylinder (Fig. 2.6*b*). A tapered surface can be produced by merely changing the angle that the directrix makes with the plane of the generatrix. When the directrix is in the plane of the circular generatrix (Fig. 2.6*c*), lines are generated which results in a plain surface when a number of generatrices and directrices are placed side by side in the direction perpendicular to the plane of the generatrix. In actual practice, the cutting is performed by a cutting edge and not a point. Thus, a series of generatrix-directrix combinations are involved and the relative motion produces a surface rather than a line.

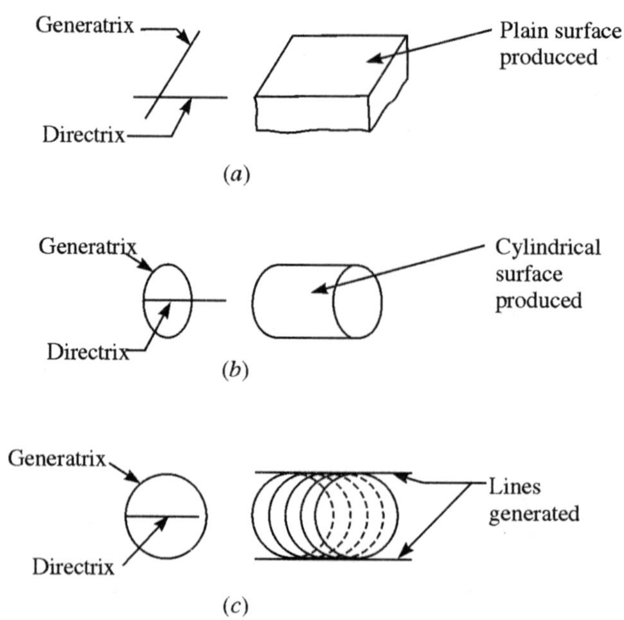

Fig. 2.6 Concept of generatrix and directrix
(*a*) Rectilinear generatrix and directrix. (*b*) Directrix perpendicular to the plane of generatrix. (*c*) Directrix in the plane of generatrix.

Basically there are two methods of producing new surfaces, the tracing method and the generation method. In the tracing method, the surface is obtained by direct tracing of the generatrices, and when the surface produced is the envelope of the generatrices the process is known as *generation*. In Figs. 2.6(*a*) and 2.6(*b*), the plane and the cylindrical surfaces are obtained by direct tracing while in Fig. 2.6(*c*) the final surface geometry is the envelope of the generatrices.

2.3 MACHINING WITH SINGLE-EDGE TOOLS

Practical cutting operations described above involve two or more cutting edges inclined at various angles. This is true for even simple turning operations. A typical lathe tool is shown in Fig. 2.7 where the principal and auxiliary cutting edges are indicated. The basic mechanism of cutting can, however, be explained by analysing the process of chip formation with single edge tools. Figure 2.8 shows a wedge-shaped tool where α is the rake angle and δ is the flank angle. Angle α defines the angle that the tool face makes with a plane normal to the cutting velocity vector and the machined surface, while δ is the angle between the flank face and the machined surface.

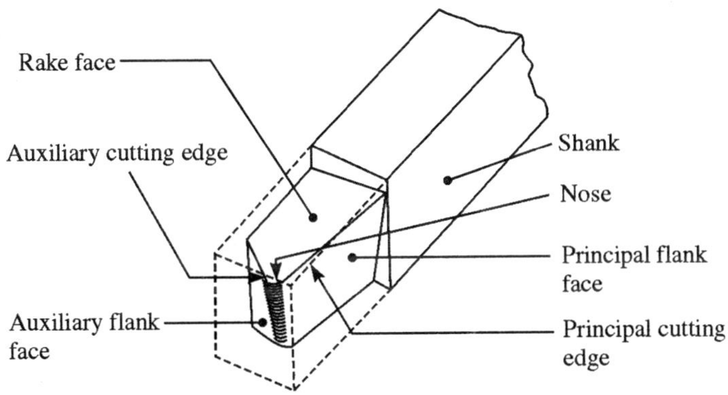

Fig. 2.7 A typical lathe tool

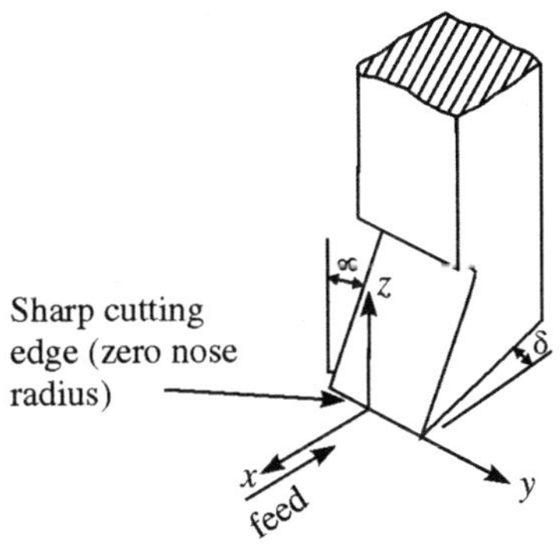

Fig. 2.8 A wedge shaped tool

The action of a wedge-shaped tool is depicted in Fig. 2.9. Here the wedge shaped tool with a straight cutting edge is constrained to move relative to the workpiece in such a way that a layer of metal is removed in the form of a chip. The cutting edge is parallel to the work surface, but in Fig. 2.9(*a*) it is inclined at an angle *i* with the normal to the cutting velocity vector *V*, *i.e.*, with the *y*-direction. In this situation, cutting takes place in three dimensions and represents the general case of oblique cutting. In Fig. 2.9(*b*), the cutting edge is perpendicular to the cutting velocity vector, *i.e.*, $i = 0$ and represents the orthogonal cutting situation, a special case of oblique cutting. Here the cutting takes place in two dimensions only and is relatively simpler to analyse theoretically.

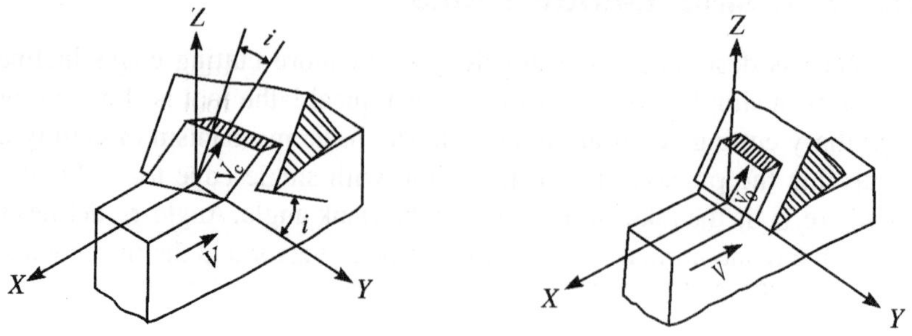

(*b*)

Fig. 2.9 (*a*) Oblique cutting (*b*) Orthogonal cutting

The basic features of these cutting operation can be summarized as follows:

2.3.1 Orthogonal Cutting

(*a*) The cutting edge of the tool is perpendicular to the direction of cutting velocity (*V*).

(*b*) The cutting edge is wider than the workpiece width and extends beyond the workpiece on either side. Also the width of the workpiece is much greater than the depth of cut. These conditions give two-dimensional flow, *i.e.*, plane strain deformation is achieved and the flow is confined to the (*x*, *z*) plane with practically no side flow.

(*c*) The chip generated flows on the rake face of the tool with chip velocity perpendicular to the cutting edge (Fig. 2.9*b*).

(*d*) The cutting forces act along *x* and *z* directions only and can be represented in the (*x*, *z*) plane.

2.3.2 Oblique Cutting

(*a*) The cutting edge of the tool is inclined at an angle *i* with the normal to the cutting velocity vector. Angle *i* is generally called the *inclination angle*.

(*b*) The chip generated flows on the rake face of the tool at an angle approximately equal to *i* with the normal to the cutting edge in the plane of the rake face (Fig. 2.9*a*).

(*c*) The cutting edge extends beyond the width of the workpiece on either side.

(*d*) The cutting forces act along all the three directions, *i.e.*, along *x*, *y* and *z* axes.

2.4 TYPES OF CHIPS

Depending on the properties of the work material and the cutting conditions, three basic types of chips are obtained in practice. These are:

(*a*) Continuous chip

(*b*) Continuous chip with built-up edge

(*c*) Discontinuous chip

2.4.1 Continuous Chip

Due to the relative motion between the tool and the workpiece, the workpiece material is compressed as it approaches the cutting edge of the tool and reaches plastic state. Plastic flow occurs and the deformed material moves upwards along the tool face and comes out as a chip. Basically the operation is one of shearing the work material in a zone extending from the tool edge (*A*) to the junction between the workpiece and the chip (*B*), as shown in Fig. 2.10. Imagine a pack of cards oriented parallel to the plane *AB*. During shearing each card of the pack is sheared through a distance just sufficient to maintain contact with the tool face. Thus, a finite thickness of the metal moves just like a card in the pack. If the cards were infinitesimally thin, the shearing across the plane will take place in a continuous manner and chip will be formed continuously.

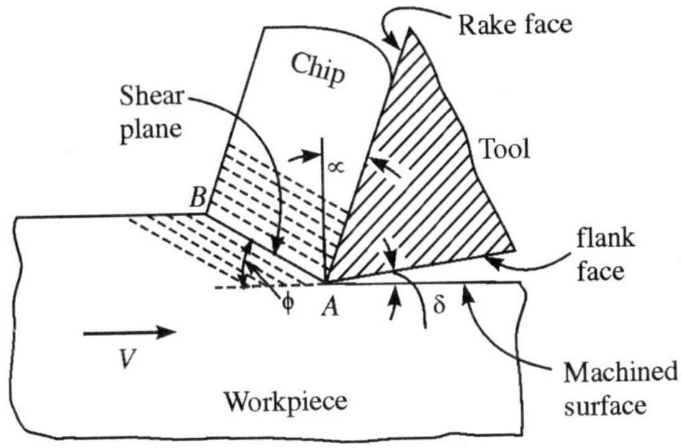

Fig. 2.10 Schematic view of idealized chip formation

This type of chip is common for most ductile materials (aluminium, copper, etc.) and is generally obtained under 'steady-state' conditions. An idealized model of continuous chip formation process is shown is Fig. 2.11(*a*). Other conditions that promote the formation of this type of chips are high cutting speed (*V*), low feed (*f*), large rake angle (α), sharp cutting tool (zero nose radius), low friction, etc.

2.4.2 Continuous Chip with Built-up Edge

Under certain conditions of temperature and pressure at the tool-chip interface, the resistance to relative motion between the chip and the tool is so severe that instead of the chip sliding over the rake face, a layer of metal is ruptured from the chip. High resistance at the chip-tool interface quickly removes the protecting layer from the rake face and this newly exposed tool surface and the freshly cut chip have affinity for each other and get welded. The presence of welded chip material further increases the resistance to flow of chip resulting in the formation of additional layers of chip material on the rake face. The process continues and the resulting pile of chip material on the tool face (built-up-edge) grows in size and then breaks when it becomes unstable and starts to pile up again (Fig. 2.11*b*). When the unstable built-up edge breaks, most of the broken pieces are carried away by the underside of the chip while some escape under the flank face of the tool resulting in poor surface finish of the machined surface. The presence of built-up edge also influences the power consumption during machining and the life of the tool. In general, low-cutting speed, high-feed and small-rake angle are conducive to the formation of built-up edge.

(*a*)

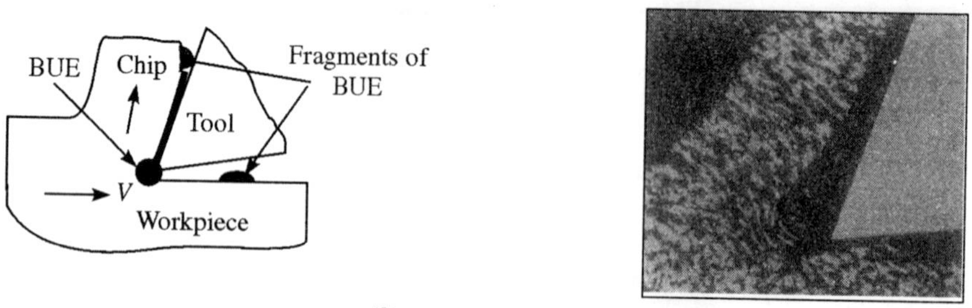

(*b*)

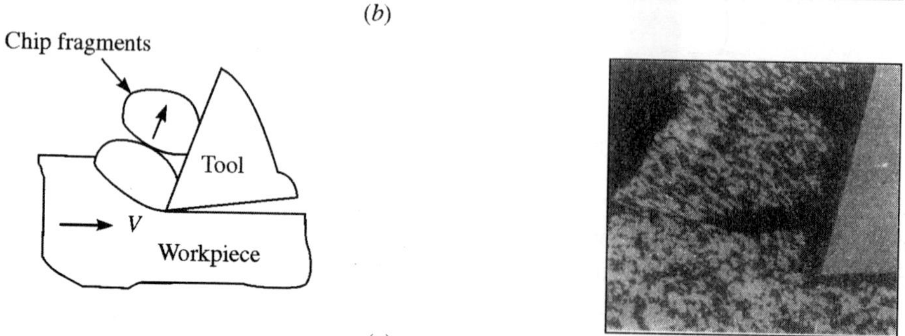

(*c*)

Fig. 2.11 Types of chips
(*a*) Continuous chip, (*b*) Continuous chip with built-up-edge, and (*c*) Discontinuous chip

The formation of built-up-edge is a transient and unstable phenomenon and occurs only in a critical speed range. At lower speeds outside the range, most materials tend to give discontinuous chips while at higher speed the formation of built-up edge may disappear. The size of the built-up edge is also influenced by speed in the critical range, where it initially increases with speed and then decreases.

The mechanism of built-up edge formation though not completely understood, it is clear that it involves welding and rupture of chip material. While welding is affected by the temperature and normal pressure near the tool tip, rupture is influenced by strain hardening and thermal softening of the chip material. During machining, the layer of workpiece material adjacent to the rake surface near the tool tip is most severely deformed and the resulting strain-hardening causes the hardness to increase significantly. As the chip slides, the bonding between the hardened material and the tool breaks at some distance inside the chip where the material is softer because of combined effect of hardness and temperature gradients. A layer of chip material thus sticks on to the tool surface near its tip and the formation of built-up edge is initiated. When the temperature in this zone is above the recrystalization temperature of the workpiece material, strain-hardening is negligible and no built-up edge is formed. Figure 2.12 indicates a typical hardness distribution while Fig. 2.13 shows the variation in temperature with speed. At speed beyond a certain value, the recrystalization temperature is reached and built-up edge formation ceases. At low speeds, the temperature is low and severe strain-hardening often leads to fracture and discontinuous chip formation even when ductile materials are machined. This behaviour is further substantiated from the surface finish versus cutting speed curve (Fig. 2.14). In region I, the surface finish is initially poor because of discontinuous chip formation but it improves as the cutting speed is increased and the discontinuous chips tend to become semi-discontinuous. In region II, built-up edge formation starts and continues till the recrystalization temperature is reached. Beyond this range, *i.e.*, in region III, continuous chips without built-up edge is obtained and surface finish improves.

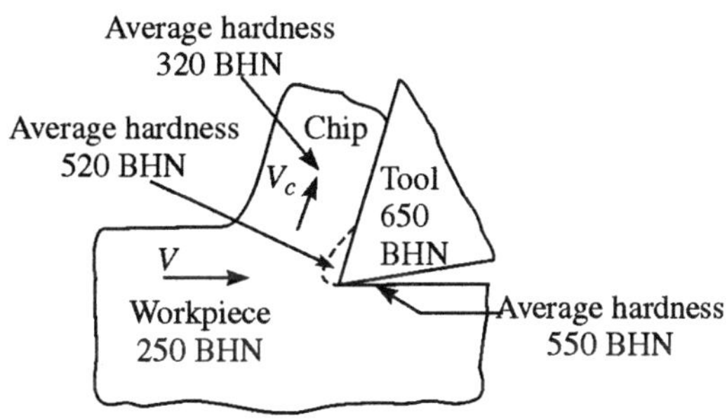

Fig. 2.12 Average hardness values in chip, built-up edge and finished surface when 250 BHN workpiece is cut by a 650 BHN tool (After Shaw, 1984)

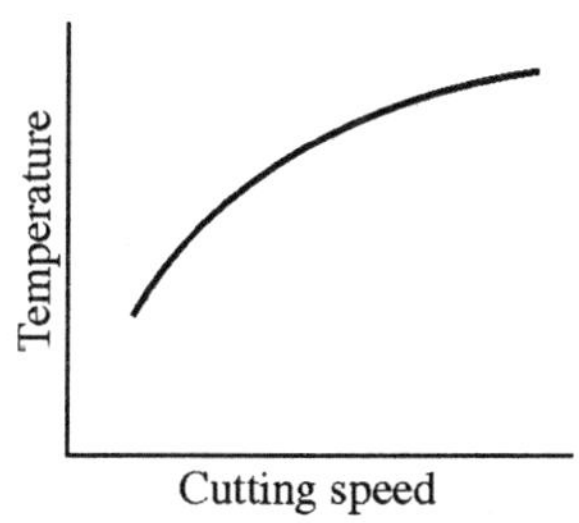

Fig. 2.13 Variation of temperature with cutting speed

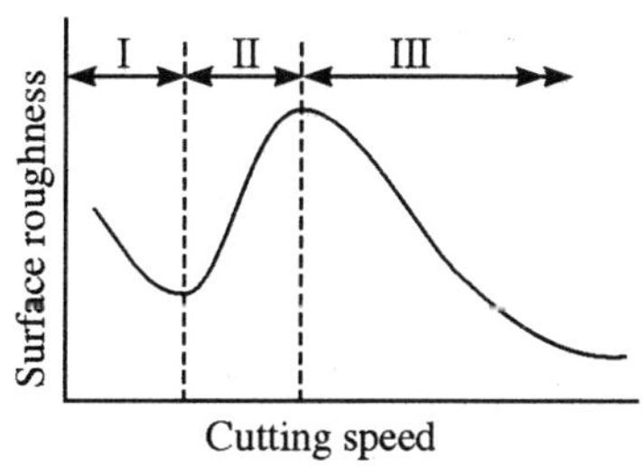

Fig. 2.14 Variation of surface roughness with cutting speed

2.4.3 Discontinuous Chip

When the work material approaches the cutting edge, it is severely compressed resulting in high strain and if the material is brittle, fracture may occur in the deformation zone itself. Thus, for brittle materials like cast iron, the chips will come out in small pieces or segments and are simply pushed away by the tool (Fig. 2.11*c*). Discontinuous chips are also obtained during machining of ductile materials when conditions are such that the material is plastically deformed beyond the critical strain. This generally happens at low speeds when the material is severely strain-hardened and the temperature is low as discussed in the preceding section. In general, low cutting speeds, high feed, and small rake angle are favourable conditions for discontinuous chip formation.

The factors that are likely to influence the formation of various types of chips can now be summarized as follows:

Factors	*Type of chip*		
	Continuous	*Continuous with B.U.E*	*Discontinuous*
Material	Ductile	Ductile	Brittle
Tool :			
Rake angle	Large	Small	Small
Cutting edge	Sharp	Dull	–
Cutting condition:			
Speed	High	Low	Low
Feed	Low	High	High
Friction	Low	High	–
Cutting fluid	Efficient	Poor	–

CHAPTER 3

Tool Geometry

3.1 SINGLE EDGE TOOLS

The action of a wedge-shaped tool was depicted in the previous chapter. The characteristic feature of this tool is present in all tools used for machining operations. It is clear that the cutting process is influenced by the inclination of the cutting edge and the orientation of the rake and flank faces. To define these orientations, a set of reference planes with a set of axes are required. Considering this simple case of a wedge-shaped tool, experience and experiments indicate that the most dominant factor influencing the cutting performance is the rake face orientation or the rake angle. The choice for the selection of reference planes is, therefore, based primarily for defining the rake angle. Accordingly, the rake angle of a wedge-shaped tool can be measured in

(*i*) a plane normal to the cutting edge and perpendicular to the plane containing the cutting edge (*OABC*),

(*ii*) a plane parallel to the cutting velocity vector and perpendicular to the plane containing the cutting edge (*ODEF*) and

(*iii*) a plane containing the cutting velocity and chip velocity vectors (*ODGH*), as shown in Fig. 3.1.

In orthogonal cutting situation these three planes are identical and will give the same value for the rake face orientation (rake angle). In oblique cutting, on the other hand, three different values for rake angle will be obtained. The rake angles corresponding to the above three planes are normal rake α_n velocity rake α_v and effective rake α_e as shown. The values for the flank angle will also have to be accordingly defined. The three angles–inclination angle, rake angle and flank angle, therefore, define the geometry of a wedge tool.

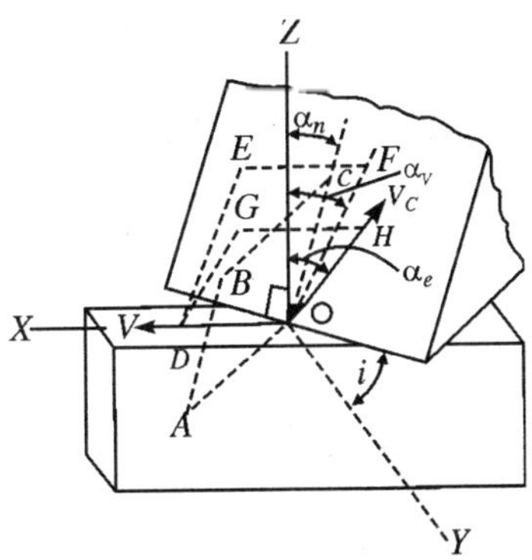

Fig. 3.1 Single edge-cutting tool

3.2 SINGLE-POINT CUTTING TOOLS

The geometry of practical cutting tools is quite complex since two or more cutting edges are involved in cutting. Metal cutting operations involving either single- or multi-point cutting tools have specified geometry. In this respect, the abrasive machining processes are quite different since the geometry of the abrasive grits (cutting points) are infinitely variable and unknown. The geometry of most cutting tools have been developed on the basis of experience rather than analysis and experiments. Since these tools had to be made in workshops, various methods of specifying the tool geometry have evolved. Some of the important standards as far as the geometry of single-point tools are concerned are:

(*i*) American system
(*ii*) Continental system
(*iii*) British system, and
(*iv*) International system

3.2.1 Reference Planes

In order to define the geometry of a single-point cutting tool, certain reference planes are necessary. Each of the above systems of specifications has a different set of planes as reference planes. Some of the planes are common in these systems. We will classify these sets in terms of the following reference systems:

1. Co-ordinate system (ASA)
2. Orthogonal system (ORS)
3. Maximum rake system (MRS)
4. Normal rake system (NRS)

The co-ordinate system of planes for a single-point turning tool are shown in Fig. 3.2(*a*). In this system, a plane parallel to the ground (horizontal plane) containing the tool shank is taken as the datum and is called the *base plane* (π_r). In other words, if the cutting velocity vector is taken as datum, the base plane is perpendicular to this velocity vector. The second reference plane, that is, the longitudinal plane π_x is along the longitudinal feed direction (primary motion) and perpendicular to the base plane. The third reference plane or the transverse plane π_y is perpendicular to both the above planes.

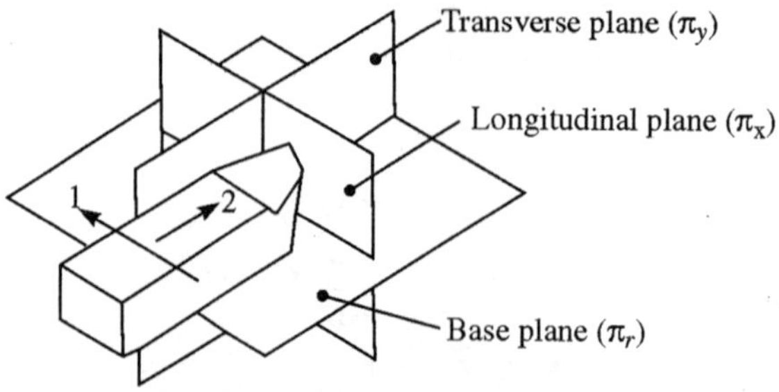

Fig. 3.2(*a*) Reference planes in co-ordinate system
(1) Longitudinal and (2) Transverse feed directions

In the orthogonal system, the reference planes are chosen from the considerations of tool operating on the workpiece on the machine tools. Again the reference planes are three mutually perpendicular planes but related to the position of the cutting edge. For a single-point turning tool (Fig. 3.2*b*) the base plane π_r is same as in the previous case but the second reference plane, the cutting plane π_c, contains the cutting edge and is perpendicular to the base plane. The third reference plane, the orthogonal plane π_o, is now perpendicular to these two planes.

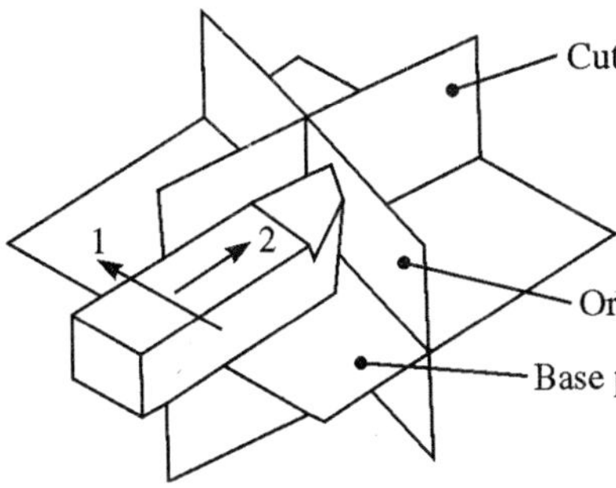

Fig. 3.2(*b*) Reference planes in orthogonal system
(1) Longitudinal (2) Transverse feed directions

In the maximum rake system Fig. 3.2(*c*), the base and the cutting planes are the same as in the orthogonal system but the third plane, the maximum rake plane π_m, is normal to the base plane π_r and is in the direction of the maximum slope of the rake face. The third plane called the *maximum rake plane* is inclined at an angle, say θ_m, with the transverse feed direction.

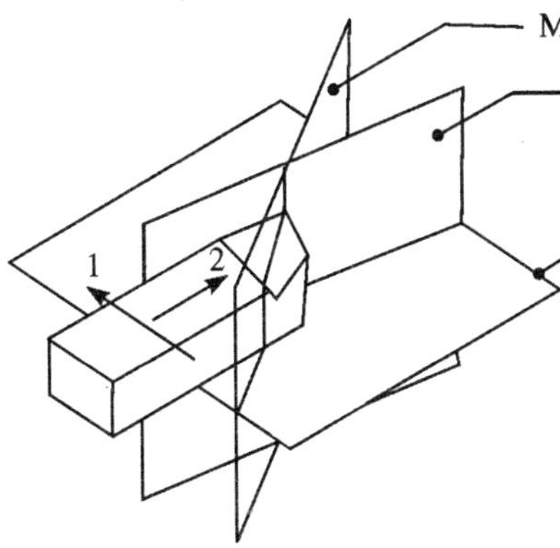

Fig. 3.2 (*c*) Reference planes in maximum rake systems
(1) Longitudinal (2) Transverse feed directions

In the normal rake system also, the base and the cutting planes are the same as in the orthogonal system but the third plane is perpendicular to the cutting edge of the tool (Fig. 3.2 *d*). The third reference plane is called the *normal plane* π_n.

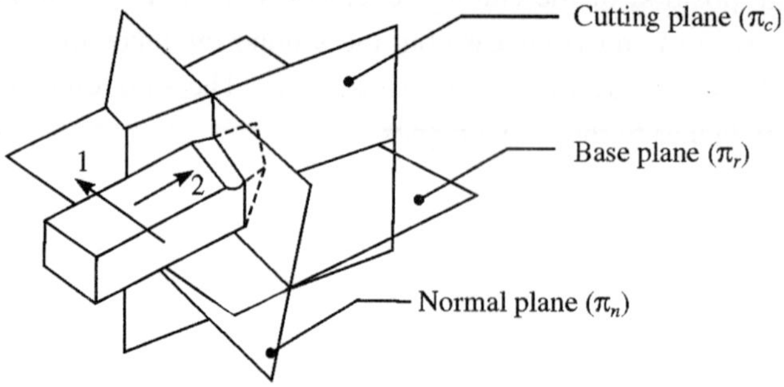

Fig. 3.2 (*d*) Reference planes in normal rake systems
(1) Longitudinal (2) Transverse feed directions

3.2.2 System of Axes

The cutting edges of a straight-turning tool in all the four reference systems are located with respect to the longitudinal and transverse feed directions, X_f and Y_f respectively. The orientation of rake and flank faces are defined with respect to the system of axes (x, y, z) as shown in Fig. 3.3 for the case of a straight-turning tool. Here subscripts *a, o, m* and *n* refer to american, orthogonal, maximum rake and normal rake systems, respectively. These are not the cartesian co-ordinate axes but traces of the reference planes. The system of axes (x_{au}, y_{au}, z_{au}) are also shown and are used for defining the auxiliary cutting edge and auxiliary flank angles in *ORS*, *MRS* and *NRS* specifications.

3.3 TOOL SPECIFICATIONS

3.3.1 American System (ASA)

The American System of specification uses the co-ordinate system of planes (Fig. 3.2*a*) and the system of axes as shown in Fig. 3.3. Consider the general shape of a single-point cutting tool (Fig. 2.7). The various angles of this tool are shown in Fig. 3.4(*a*). In american system, the orientation of the rake face is defined by two rake angles, the back rake α_b and the side rake α_s. Similarly, the end flank and the side flank face orientations are defined by the end flank angle δ_e and the side flank angle δ_s and the end and the side cutting-edges are defined by the end and the side cutting-edge angles γ_e and γ_s, respectively. The nose radius of the tool tip is r.

The back rake (α_b) is defined in the transverse plane π_y and is the angle made by the rake face with respect to y_a-axis (Figs. 3.2*a*, 3.3 and 3.4*a*). Similarly, the side rake angle α_s is defined in the longitudinal plane π_x and is the angle made by the rake face with respect to x_a-axis. The side flank angle δ_s is the angle by which the principal flank surface of the tool wedge clears the velocity axis z_a in the longitudinal plane π_x and the end flank angle δ_e is the angle by which the same point on the tool wedge clears the z_a-axis on the transverse plane π_y. Similarly, the auxiliary flank surface is also relieved from the velocity axis z_a in π_x and π_y planes. The end cutting-edge angle γ_e is located with respect to x_f-axis while the side cutting edge angle γ_s is located with respect to y_f-axis in the base plane π_r.

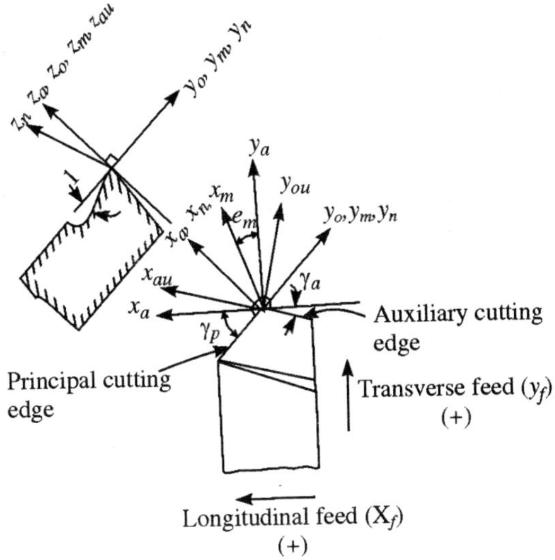

Fig. 3.3 System of axes

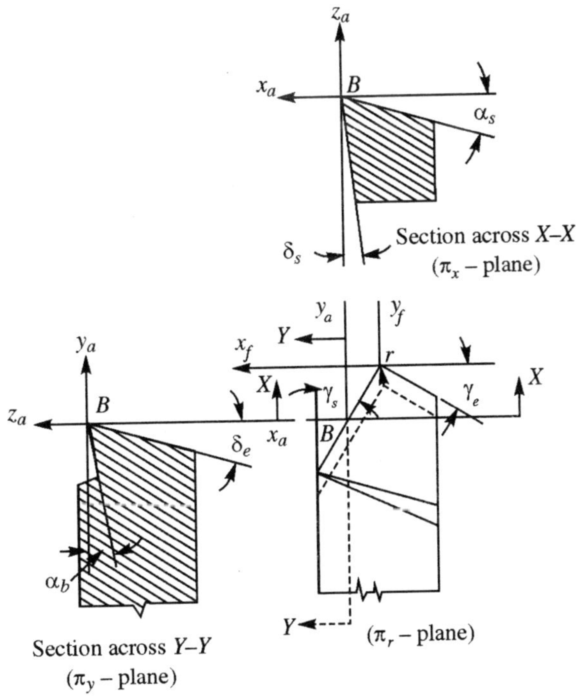

Fig. 3.4(*a*) Single-point tool specification in American system (*ASA*)

The tool geometry specified in this system of specification has no direct relation to the mechanics of cutting. From tool grinding point of view, this system of specification is convenient because of ease in determining the reference co-ordinates. Though the tool grinding procedure is simple in *ASA* specification it is inaccurate unless corrections are made in the tool angles while grinding.

The various angles in this system are specified in a special sequence as

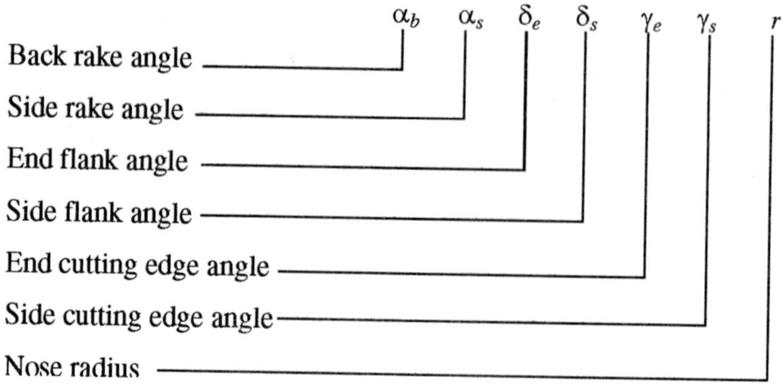

Thus, a tool geometry 8-8-5-7-10-6-2 would mean that a tool has a back rake angle of 8°, side rake angle of 8°, end flank angle of 5°, side flank angle of 7°, end cutting-edge angle of 10°, side cutting-edge angle of 6° and a nose radius of 2 mm.

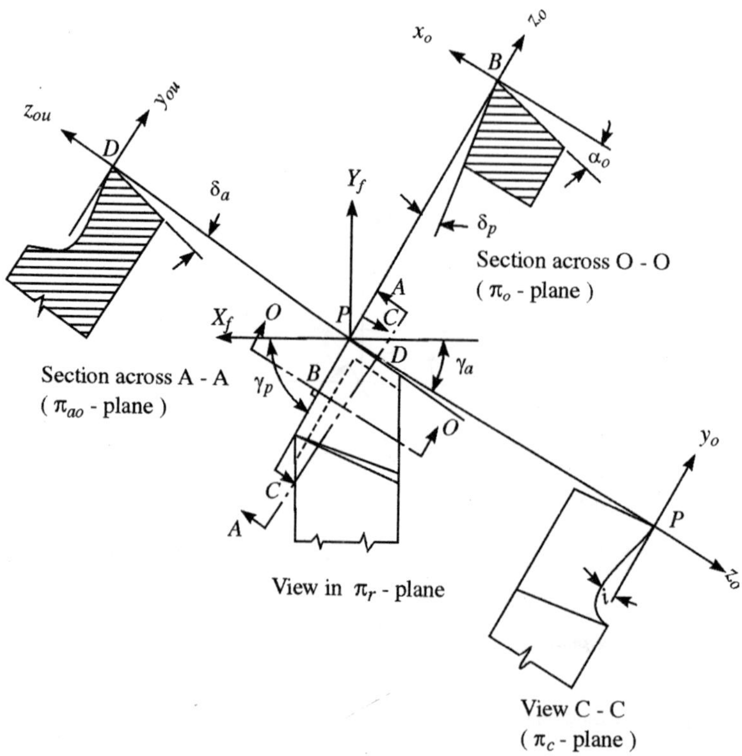

Fig. 3.4(*b*) Single-point tool specification in Continental system (ORS)

3.3.2 Continental System (ORS)

This system includes the German (DIN), Russian and Czech systems. They all use the orthogonal system of reference planes shown in Fig. 3.2(*b*) and set of reference axes shown in Fig. 3.3. The various angles of the turning tool in this system are shown in Fig. 3.4(*b*) against appropriate projections. The rake face is now defined in terms of the orthogonal rake angle α_o and inclination angle *i*, while the principal and auxiliary flanks are defined by angles δ_p and δ_a. Similarly, the principal cutting-edge angle is γ_p and the auxiliary cutting-edge angle is γ_a and *r*, as before, is the nose radius.

The orthogonal rake α_o is defined in the orthogonal plane π_o and is the angle made by the rake face of the tool with the x_o-axis, while the inclination angle *i* is the angle made by the principal cutting-edge with y_o-axis in the cutting plane π_c (Figs. 3.2*b*, 3.3 and 3.4*b*). The principal flank angle δ_p is the angle that the principal flank surface of the tool wedge makes with the tangent at the cutting edge along the z_o-axis in the π_o plane. Similarly, in an auxiliary orthogonal plane π_{ao}, that is, a plane perpendicular to the plane containing the auxiliary cutting-edge, the auxiliary flank angle δ_a is defined. The principal cutting-edge angle γ_p and the auxiliary cutting-edge angle γ_a are defined in π_r plane. γ_p and γ_a are the angles made by the principal and auxiliary cutting-edges of the tool with respect to the X_f-axis.

The orthogonal system of specification is related to the behaviour of cutting tools since the geometry of chip formation is always analysed with respect to the cutting-edge and here the plane containing the cutting-edge has been chosen as one of the reference planes. Also all angles, except the orthogonal rake angle, are true angles. The grinding of the tool, however, requires necessary corrections for inclination effect of the cutting-edge.

The sequence of angles for specification in this system is

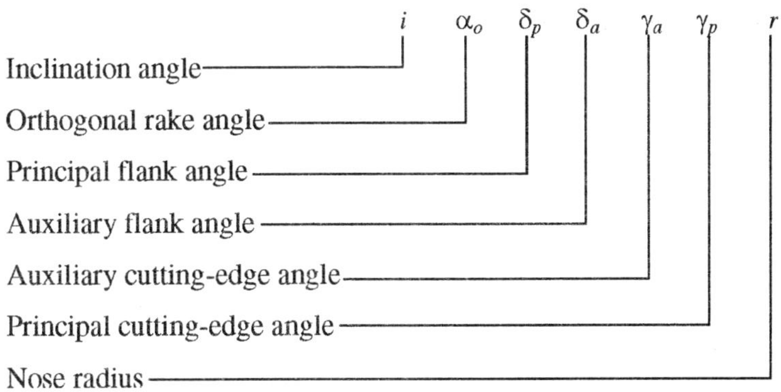

3.3.3 British System (MRS)

The British System of specification uses the reference planes of the maximum rake system (Fig. 3.2*c*) and the system of axes shown in Fig. 3.3. In this system, the rake face orientation is defined in terms of maximum rake angle α_m and the maximum rake plane angle θ_m as shown in Fig. 3.4(*c*). The maximum rake angle α_m is the angle made by the rake face in the oblique plane π_m with the x_m-axis and θ_m is the angle of the oblique plane with the y_f-axis in π_r plane. The concept of maximum rake plane was introduced since it was thought to contain the chip velocity vector. The other angles are the same as in the continental system.

The British system of specification of tool geometry is considered to be unrelated to the mechanics of cutting. Grinding and inspection of the tool in this system are, however, easy.

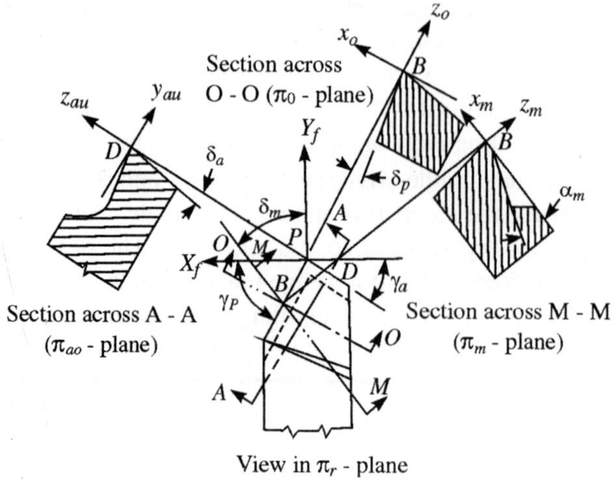

Fig. 3.4(*c*) Single-point tool specification in British system (MRS)

The sequence of angles in this system of specification is

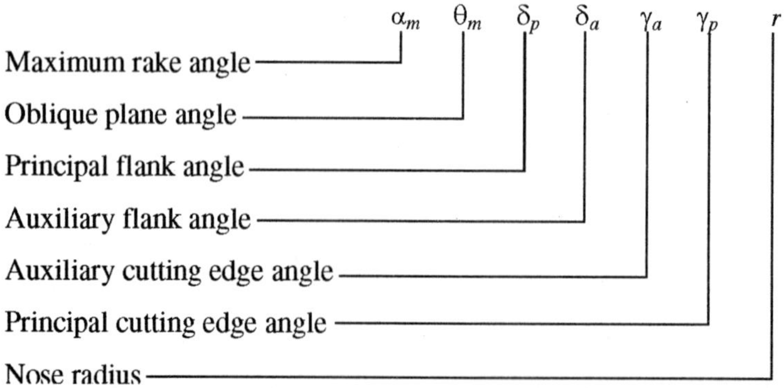

3.3.4 International System (NRS)

It is clear from the foregoing sections that the system of specification can be quite complex in the absence of standardization concerning definitions of various tool angles. The American, Continental and British systems define tool angles with reference to different sets of planes and axes. The International system (ISO) in this regard is an attempt to synthesize the different ways of representing tool nomenclature. The Indian Specification (IS) which was earlier based on DIN standards now follows the ISO system. The system suggests the use of normal rake system of reference planes shown in Fig. 3.2(*d*) with the set of axes shown in Fig. 3.3. In this system of specification, except the normal rake, the principal normal flank and the auxiliary normal flank angles, the remaining tool angles are defined in the same manner as in the Continental system. The normal rake angle α_n and the principal normal flank angle δ_{pn} are measured in the normal plane π_n with reference to the axes x_n and z_n, respectively.

The auxiliary normal flank angle δ_{an} is measured with respect to z_{an} axis in the auxiliary normal plane π_{an}. Orientation of rake and flank faces as well as the cutting edges are shown in Fig. 3.4(*d*). It may be mentioned that with this system of planes and axes, all angles referred to are true angles and tool geometry is related to the mechanics of cutting. Tool grinding is easy and the angles do not require any compensation or correction while grinding. Both the normal and the orthogonal systems will give the same values for the various tool angles for zero inclination angle (i = 0).

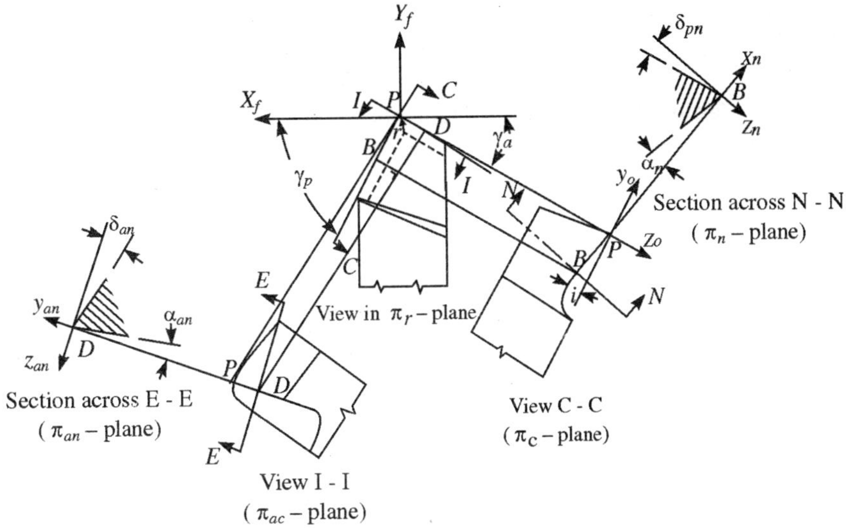

Fig. 3.4(*d*) Single-point tool specification in International system (NRS)

The sequence of angles in the international specification is

Inclination angle	i
Orthogonal rake angle	α_n
Principal flank angle	δ_{pn}
Auxiliary flank angle	δ_{an}
Auxiliary cutting-edge angle	γ_a
Principal cutting-edge angle	γ_p
Nose radius	r

In Fig. 3.5 the tool angles in American, Continental, British and International systems are shown together for comparison purposes. Table 3.1 summarizes the nomenclature systems described for single-point tools while the sequence of tool angles in different systems are listed together in Table 3.2.

Table 3.1 Different Nomenclature Systems for Single-point Tools

Tool Section	*American System*	*Continental System*	*British System*	*International System*
Rake face	α_b, α_s	i, α_o	α_m, θ_m	i, α_n
Flank face	δ_e, δ_s	δ_a, δ_p	δ_a, δ_p	δ_{an}, δ_{pn}
Cutting edge	γ_e, γ_s	γ_a, γ_p	γ_a, γ_p	γ_a, γ_p
Nose radius	r	r	r	r

Table 3.2 Sequence of Tool Angles in Different Systems

Systems	*Specifications*
American (ASA)	$\alpha_b - \alpha_s - \delta_e - \delta_s - \gamma_e - \gamma_s - r$
Continental (ORS)	$i - \alpha_o - \delta_p - \delta_a - \gamma_a - \gamma_p - r$
British (MRS)	$\alpha_m - \theta_m - \delta_p - \delta_a - \gamma_a - \gamma_p - r$
International (NRS)	$i - \alpha_n - \delta_{pn} - \delta_{an} - \gamma_a - \gamma_p - r$

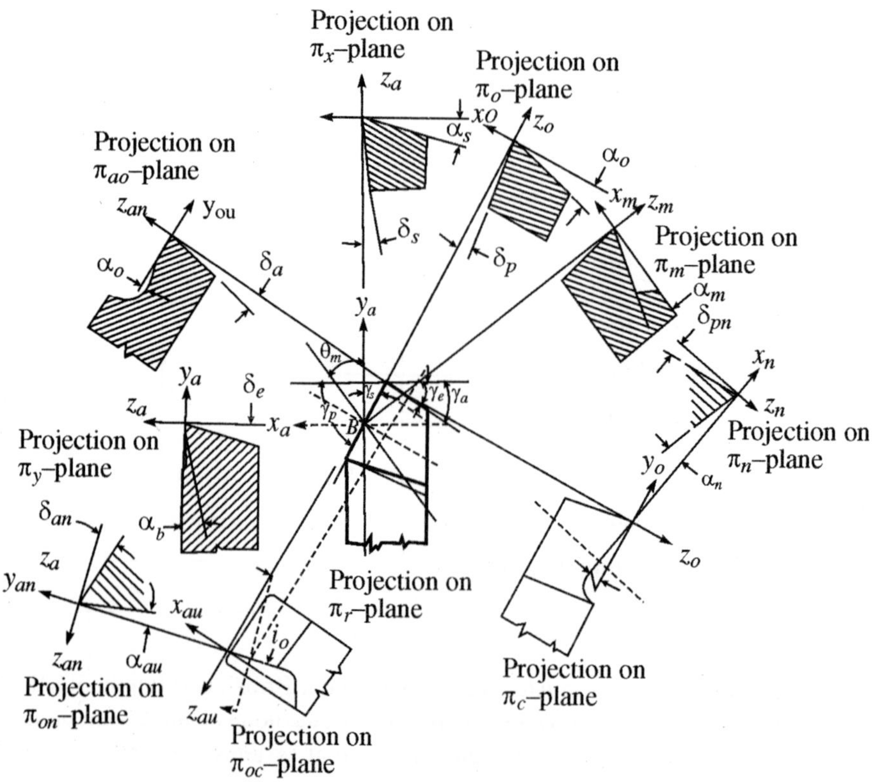

Fig. 3.5 Single-point tool specification in American, Continental, British and International systems

3.3.5 Sign Convention

Figure 3.6 shows how positive and negative angles are denoted. The rake angle is described in the orthogonal plane while the inclination angle is shown in the cutting plane (ORS system). In the left hand cork screw rule, when the axis x_o is rotated towards z_o-axis to meet the rake face, the rake angle is taken as positive and when the rotation is away from the z_o-axis, the angle is taken as negative as shown in the figure. Similarly, the inclination angle i is taken as positive when y_o is rotated towards z_o to meet the cutting edge and is negative when the rotation is away from z_o. The left hand cork screw sign convention applies to other systems also but appropriate axes must be rotated to meet the tool surface.

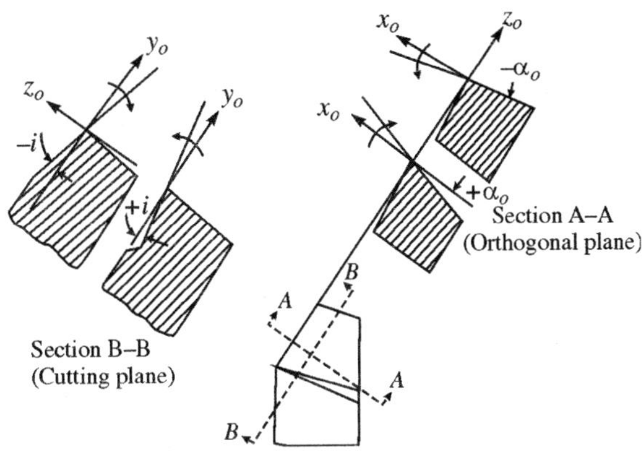

Fig. 3.6 Sign convention for rake and inclination angles

3.4 CONVERSION OF TOOL ANGLES

Conversion of tool angles from one system to the other can be obtained by transforming the projections from the respective reference planes. This is illustrated by means of an example of conversion of American System (ASA) to Continental System (ORS). Figure 3.7 shows the projection of tool angles on relevant reference planes. Points A, T and Q on the base plane (π_r) are transformed on π_y plane as A_y, T_y and Q_y and are further transformed on π_x and π_o planes. Similarly, the cutting edge on π_c plane which is inclined at an angle i is transformed on π_o plane. Points B_c and B_c' on π_c plane now become B_o and B_o' on π_o plane and point B_x on π_x plane corresponds to B_o on π_o plane. Point T_x on the rake face in π_x plane is transformed to corresponding point T_y on the rake face in π_y plane. The inclination angle i can now be evaluated from

$$\tan i = \frac{B_c B_c'}{A_c B_c'}$$

$$= \frac{T_x T_x'' - T_x T_x'}{AB}$$

$$= \frac{T_y T_y'' - T_x T_x'}{AB}$$

$$= \frac{A'_y T''_y \times \tan\alpha_b - T'_x B_x \times \tan\alpha_s}{AB}$$

$$= \frac{AT}{AB}\tan\alpha_b - \frac{BT}{AB}\tan\alpha_s$$

or

$$\tan i = \sin\gamma_p \tan\alpha_b - \cos\gamma_p \tan\alpha_s \qquad ...(3.1)$$

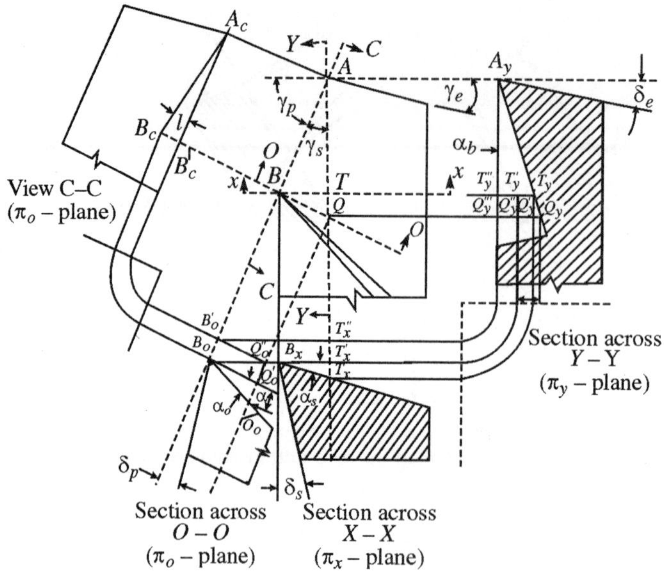

Fig. 3.7 Tool-angle relationship in ORS and ASA systems

Similarly, the orthogonal rake α_o can be evaluated from

$$\tan\alpha_o = \frac{Q_o Q'_o}{B_o Q'_o}$$

The point Q''_y and Q_y in π_y plane correspond to points Q'_o and Q_o in π_o plane, *i.e.*, $Q''_y\ Qy = a = Q_oQ'_o$. Thus,

$$\tan\alpha_o = \frac{Q'_y Q''_y + Q_y Q'_y}{B_o Q'_o}$$

$$= \frac{T_x T'_x + Q_y Q'_y}{BQ}$$

$$= \frac{B_x T'_x \tan\alpha_s + Q'_y T_y \tan\alpha_b}{BQ}$$

$$= \frac{BT}{BQ}\tan\alpha_s + \frac{QT}{BQ}\tan\alpha_b$$

or

$$\tan\ \alpha_o\ =\ \sin\gamma_p \tan\alpha_s + \cos\gamma_p \tan\alpha_b \qquad ...(3.2)$$

The principal cutting edge angle γ_p is simply equal to $(90 - \gamma_s)$.

For obtaining the tool angles in the International System (NRS), the normal plane is taken as the reference plane in which the rake angle α_n is measured. This plane is inclined to the orthogonal plane (π_o) by an angle i. Hence, the point Q_o in the π_o plane is rotated by an angle i and its projection Q_n is obtained as shown in Fig. 3.8. The rake angle α_n can be evaluated from

$$\tan\ \alpha_n\ =\ \frac{Q'_oQ_n}{Q'_oB_o}$$

$$= \frac{Q_oQ'_o}{Q'_oB_o}\cos\ i$$

or

$$\tan\ \alpha_n\ =\ \cos\ i\ \tan\ \alpha_o \qquad ...(3.3)$$

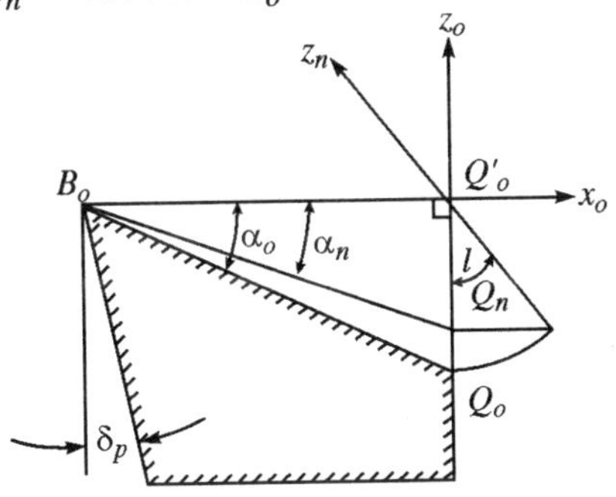

Selection in π_o–plane

Fig. 3.8 Tool-angle relationship in ORS and NRS

The geometric equations required for converting tool angles from other systems into International system can be obtained in a similar manner. A consolidated set of equations for converting various tool angles are given below :

American System:

$$\tan\ i = \tan\alpha_b \cos\gamma_s - \tan\alpha_s \sin\gamma_s \qquad ...(3.4)$$

$$\tan\ \alpha_n\ =\ \cos\ i\ (\tan\alpha_s \cos\gamma_s + \tan\alpha_b \sin\gamma_s) \qquad ...(3.5)$$

$$\gamma_p\ =\ 90 - \gamma_s \qquad ...(3.6)$$

$$\cot^2\delta_p\ =\ \cot^2\delta_e + \cot^2\delta_s - \tan^2 i \qquad ...(3.7)$$

$$\cot\delta_{pn}\ =\ \cos\ i\ \cot\ \delta_p \qquad ...(3.8)$$

Continental System:

$$\tan\ \alpha_n\ =\ \cos\ i\ \tan\ \alpha_o \qquad ...(3.9)$$

$$\cot\delta_{pn}\ =\ \cos\ i\ \cot\ \delta_p \qquad ...(3.10)$$

British System:

$$\tan i = \tan\alpha_m \cos(\gamma_p + \theta_m) \quad ...(3.11)$$

$$\tan\alpha_n = \tan(\gamma_p + \theta_m)\sin i \quad ...(3.12)$$

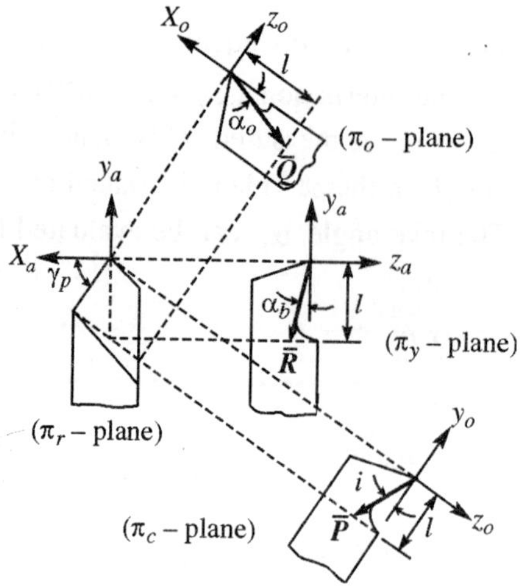

Fig. 3.9 Rake angle relationship in ASA and ORS systems

There are several other methods of obtaining the angle relationships between various systems. One such method, the vector method, is illustrated here for obtaining the rake angle relationship between ASA and ORS systems.

The relevant views of a single point tool are shown in Fig. 3.9. Consider vectors *P*, *Q* and *R* on the rake face of the tool with *P* along the cutting edge (shown in the π_c -plane), *Q* at right angles to the cutting edge (shown in π_o -plane) and *R* in plane π_y. Taking *i*, *j* and *k* as unit vectors along x_a, y_a and z_a directions, the vectors *P*, *Q* and *R* can be written as

$$P = i\cos\gamma_p - j\sin\gamma_p - k\tan i \quad ...(3.13)$$

$$Q = i\sin\gamma_p - j\cos\gamma_p - k\tan\alpha_o$$

$$R = j - k\tan\alpha_b$$

Since the vectors lie in one plane, their mixed product is zero. Therefore,

$$\begin{pmatrix} \cos\gamma_p & -\sin\gamma_p & -\tan i \\ -\sin\gamma_p & -\cos\gamma_p & -\tan\alpha_o \\ 0 & -1 & -\tan\alpha_b \end{pmatrix} = 0 \quad ...(3.14)$$

and its solution gives

$$\tan\alpha_b = \sin\gamma_p \tan i + \cos\gamma_p \tan\alpha_o \quad ...(3.15)$$

Other relationships can be similarly obtained by taking vectors along appropriate directions and in relevant planes.

3.5 SELECTION OF TOOL ANGLES

The optimum value of any tool angle depends primarily on the tool-work combination and the output parameters such as cutting forces, surface finish, heat generation, etc. How these considerations are brought about is discussed briefly in the following sub-sections.

3.5.1 Rake Angles

Single-point cutting tools (Fig. 2.7) are designed to cut on the principal cutting edge, the nose radius and to some extent on the auxiliary cutting edge and the chip generated flows on the rake face. The rake face controls the directions of the resultant force on the tool and chip flow. With zero inclination angle or zero back rake angle the chip will flow parallel to the work surface (see Fig. 2.9) and may cause removal problem. It may even entangle with the work surface on curling and cause damage. With appropriate inclination angle or back rake angle the chip can be made to flow away from the workpiece and strike a suitable chip breaker, curl and break into small fragments for easy disposal. The rake angle $(\alpha_s, \alpha_o, \alpha_m, \text{or } \alpha_n)$ influences the cutting forces, power and surface finish. The larger the rake angle, the lower are the cutting forces and power and better is the surface finish. Large rake, however, decreases the cutting angle between the rake and principal flank faces and less metal is available to support the tool and conduct the heat generated due to plastic deformation and friction. The angle chosen is, therefore, a compromise between efficient cutting and tool strength. Further, the angle chosen could be positive or negative depending upon the desired tool strength. For rough machining, often carbide tools with negative rakes are used. On a negative rake tool, the initial impact of cutting occurs away from the cutting edge and its edge has added strength.

The rake angle $(\alpha_s, \alpha_o, \alpha_m, \text{or } \alpha_n)$ for single-point HSS tools generally vary between 5° and 15° degrees depending upon the work material, higher values being used for softer materials like aluminium. For steel and cast iron the range may be 8° – 12° and 5° – 10° degrees, respectively, depending upon the hardness. In general, harder is the work material, the lower is the rake angle.

3.5.2 Flank Angles

Flank angles $(\delta_e, \delta_s, \delta_p, \delta_a, \delta_{pn} \text{ or } \delta_{an})$ are provided to enable the sides of the tool to clear the workpiece and not to rub against it during cutting. For example, during straight turning, the principal flank angle δ_p must be greater than the helix angle of the cut. The larger the angle, the lower is the flank face wear but this would weaken the tool and also affect conductivity. Thus, in general, the flank angles should be as large as possible and still provide enough support to the cutting edge. The angle depends upon the tool and work material. For HSS and carbide tools the values of these angles are in the range of 5°– 12° and 5°– 8° degrees, respectively. Clearly, higher values are used for tougher tool materials and lower ones for brittle tool materials. Similarly, higher flank angles may be used for machining of softer materials and lower ones for harder materials. Interrupted cutting requires additional tool strength due to impact loading, therefore lower flank angles are used. Flank angles have only marginal effect on cutting forces, power and surface finish. Therefore, as long as the angle is large enough to avoid rubbing, the tool performs satisfactorily.

3.5.3 Cutting-Edge Angles

The auxiliary or end cutting-edge angle $(\gamma_a \text{ or } \gamma_e)$ is provided to clear the cutting edge from the machined surface and reduce tool chatter. Too large an angle, however, weakens the tool and affects heat conduction. The principal or side cutting-edge angle (γ_p or γ_s) affects tool life as well as surface finish. With principal or side cutting-edge angle the tool first contacts the workpiece some distance away from the tip and also increases the tool length in action during cutting. The edge, therefore, lasts longer and conductivity also increases. At large principal cutting-edge angles, the force component trying to separate the tool from the workpiece increases and promotes chatter. Large angle also weakens the tool. For most machining operations an angle in the range of 5°-15° degrees for both cutting edges has been found to be quite satisfactory.

A sharp tool tip is highly stressed, has short life and results in feed marks on machined surface. The nose radius smoothens these marks, strengthens the tool and also increases the effective conductivity of the cutting tool. The nose radius, however, increases the forces and tool chatter. In single-point tools, a nose radius in the range of 0.5 – 3 mm has been found to be quite satisfactory.

Typical tool angles in ASA specifications for single point HSS and cemented carbides turning tools are given in Table 3.3.

Table 3.3 Typical Tool Angles in ASA Specifications for Single-Point Turning Tools

Workpiece material	*Tool material*	*Tool angles (degrees)*					
		α_b	α_s	δ_e	δ_s	γ_e	γ_s
Mild steel	1. HSS	8	10	5	5	15	15
	2. Cemented carbide						
	(*i*) Brazed	0	6	5	5	15	15
	(*ii*) Throw-away	–5	–5	5	5	15	15
Cast iron	1. HSS	5	8	5	5	15	15
	2. Cemented carbide						
	(*i*) Brazed	0	6	5	5	15	15
	(*ii*) Throw-away	–5	–5	5	5	15	15
Aluminium alloys	1. HSS	20	15	12	10	5	5
	2. Cemented carbide						
	(*i*) Brazed	3	15	5	5	15	15
	(*ii*) Throw-away	0	5	5	5	15	15
Stainless steel	1. HSS	0	10	5	5	15	15
	2. Cemented carbide						
	(*i*) Brazed	0	6	5	5	15	15
	(*ii*) Throw away	–5	–5	5	5	15	15
Tool steel	1. HSS	–3	10	5	5	15	15
	2. Cemented carbide						
	(*i*) Brazed	0	–5	5	5	15	15
	(*ii*) Throw away	–5	–5	5	5	15	15

3.6 MULTI-POINT CUTTING TOOLS

3.6.1 Geometry of Milling Cutter

There are a large variety of milling cutters in use. The geometry of peripheral milling cutters can, however, be compared directly with single-point cutting tools discussed earlier. Its teeth act like single-point tools and have cutting edges and angles similar to single-point tools. It is for this reason, a typical helical milling cutter has been chosen to describe the geometry of milling cutters. The various surfaces and angles of such a cutter are shown in Fig. 3.10.

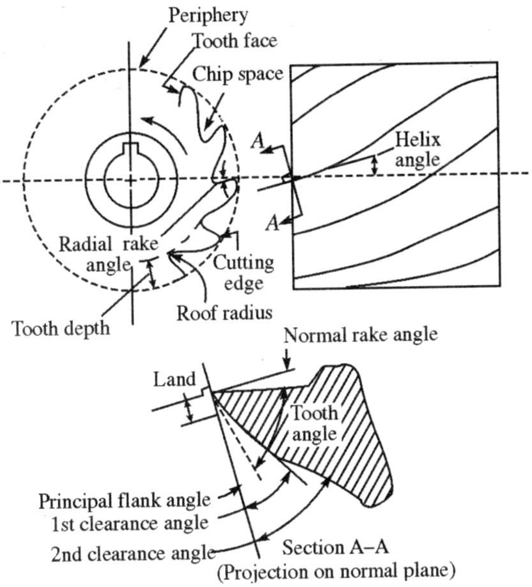

Fig. 3.10 Geometry of helical milling cutter

The cutter has a helical cutting edge and is equivalent to an oblique tool (Fig. 2.9a) having an inclination angle i equal to the helix angle ψ of the cutter. In plain milling cutters, the helix angle is zero, *i.e.*, the cutting edges are straight and is equivalent to an orthogonal tool (Fig. 2.9b). The radial rake angle α_r of these cutters are the same as the back rake angle of single-point tools and is equal to the orthogonal rake because principal cutting edge angle $\gamma_p = 0$. A section normal to the helical cutting edge gives the normal rake angle α_n. It is clear that these angles are related ($\tan\alpha_n = \tan\alpha_r \cos\psi$).

Helical cutting edges are proved to obtain smoother cuts since a helical cutter remains in contact for a larger time than a plain milling cutter for the same cutting conditions. In plain cutters, the forces on the cutting edge rise from zero to a maximum value during tool engagement and then fall to zero level during disengagement. In helical cutters, on the other hand, only a part of the cutting edge is engaged in cutting at any time. Also, depending upon the teeth spacing, two or more edges may cut simultaneously. These factors reduce the fluctuations in the spindle torque and provide smoother cutting. Similar effect can also be achieved by increasing the number of teeth. The requirement of space in between the teeth to accommodate the chips generated, however, puts a limit on the number of teeth on a cutter. The first and second clearance angles are essentially provided to increase this chip space. This aspect is discussed in detail in Section 10.4.2.

The rake and flank angles are fixed from the same considerations as those for single-point tools. Negative rake angles are common for cemented carbides when machining steel because they increase tool strength and reduce shock on the cutting edge. Typical value of radial rake angle and radial flank angle are shown in Table 3.4.

Table 3.4 Typical angles for side milling cutters

Workpiece material	*Tool material material*	*Radial rake angle (deg.)*	*Radial flank angle (deg.)*
Mild steel	1. HSS 2. Cemented carbide	10 0	6 6
Cast iron	1. HSS 2. Cemented carbide	12 5	5 6
Aluminium alloys	1. HSS 2. Cemented carbide	15 10	8 8
Stainless steel	1. HSS 2. Cemented carbide	10 0	5 6
Tool steel	1. HSS 2. Cementred carbide	10 5	5 6

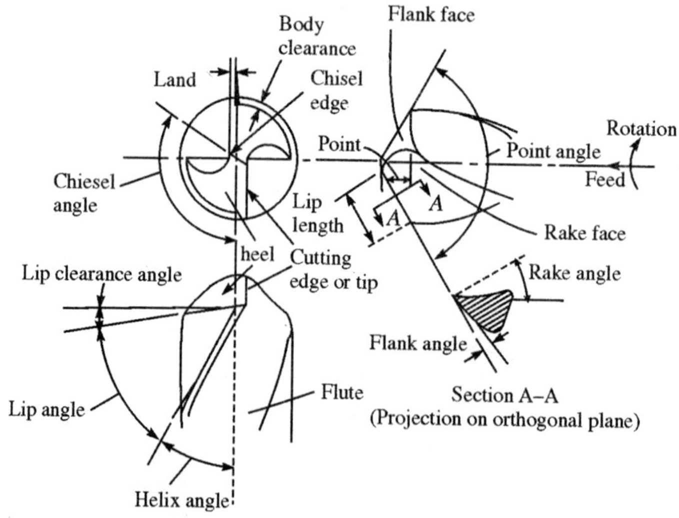

Fig. 3.11 Geometry of twist drill

3.6.2 Geometry of Twist Drill

A common form of twist drill is shown in Fig. 3.11. Cutting occurs on the chisel edge and the cutting tips and the chips generated travel up the axis of the drill through the flute. The cutting tips here correspond to the cutting edge of single-point tools. A section perpendicular to the cutting tip clearly indicates the rake and flank angles. The rake angle, however, varies along the cutting edge and is largest at the periphery. This angle should, therefore, be defined in relation to its location. The rake results primarily from the helix angle of the flute. The average value of helix angle is around 30 degrees for machining of ferrous metals and 45° for non-ferrous metals. The flank angle must be larger than the lead angle of the path of the cutting-edge to prevent rubbing. Its value generally varies between 10° and 15° at the outer periphery. No cutting occurs on the drill periphery except very near the outer corners which may be considered as the auxiliary cutting-edge. The diameter of the drill is decreased over a certain portion of its circumference leaving the land at the full diameter to support the drill against the hole. This reduces rubbing between the drill and the hole and also allows the cutting fluid to reach the drill point. The drill diameter is also decreased along its length to reduce rubbing. The point angle between the cutting-edges provide gradual entry for the drill in a similar manner as the principal cutting-edge of a single-point tool. Its value varies between 110° and 120° for machining soft materials and 120°-140° degrees for hard materials. Typical angles for HSS twist drills are summarized in Table 3.5.

Table 3.5 Typical angles for HSS drills

Workpiece material	*Point angle (deg.)*	*Lip clearance angle (deg.)*	*Chisel angle (deg.)*	*Helix angle (deg.)*
Mild steel	118	10	130	30
Cast iron	118	10	130	30
Aluminium alloys	110	15	130	35
Stainless steel	118	8	130	30
Tool steel	130	8	130	30

CHAPTER 4

Mechanics of Orthogonal Cutting

4.1 INTRODUCTION

Basically all machining operations can be considered as either oblique or orthogonal cutting operations, the former requiring three dimensions to specify and the latter two. Since most practical cutting operations are complex, the simple case of orthogonal cutting where a straight-edge tool moves relative to the workpiece in a direction perpendicular to its cutting edge has received maximum attention. During the last 40 years or so, several models have been developed, some of them have been fairly successful in describing the metal cutting process. In this regard, the analysis presented by Merchant (1944) is still considered the most significant contribution. The case of orthogonal cutting with continuous chip formation is considered in this chapter and the theory will be developed to illustrate some of the important features of the process.

4.2 THEORY

Figure 4.1(a) shows the idealised model of continuous chip formation during orthogonal cutting. In drawing this force equilibrium diagram the simplifying assumptions that have been made are:

1. The tool tip is sharp (zero edge radius) and that the chip makes contact only with the rake face of the tool.
2. The cutting edge is perpendicular to the cutting velocity.
3. The deformation is two-dimensional, *i.e.*, no side spread.
4. The deformation takes place in a very thin zone so that the deformation zone could be reasonably represented by a plane (AB), called the *shear plane*.
5. Continuous chips without *built-up-edge* are formed.
6. The workpiece material is rigid, perfectly plastic.
7. The coefficient of friction is constant along the chip-tool interface.
8. The resultant force on the chip R' applied at the shear plane is equal, opposite and collinear to the resultant force R applied to the chip at the chip-tool interface.

4.2.1 Force Relationships

The equilibrium of forces associated with the cutting process can be obtained by considering the chip as a free body held in equilibrium by the action of the forces transmitted across the shear plane and the chip-tool interface (Fig. 4.1*a*). The resultant force R' applied on the chip at the shear plane may be related to other important forces such as the cutting force or the tangential force F_p in the direction of cutting velocity V and the thrust force or the normal force F_q perpendicular to the work surface.

Alternatively, the force R' can be resolved in terms of shear force F_S along the shear plane and normal force F_N perpendicular to the shear plane. Similarly, force R applied to the chip at the chip-tool interface can be related to the friction force F on the rake face and N, the normal force on the rake face. It is clear that for equilibrium, resultant force R' must be equal, opposite and collinear to the force R. A complete force diagram is shown is Fig. 4.1(b) where the force system is shown superimposed at the cutting edge of the tool. Here R is the diameter of the circle.

From Fig. 4.1(b),

$$F_S = R\cos(\phi + \lambda - \alpha) \quad \text{...(4.1)}$$

and

$$F_S = k A_s \quad \text{...(4.2)}$$

where ϕ is the angle of inclination of the shear plane AB to the direction of cutting, called the *shear angle*, λ is the mean angle of friction between chip and tool ($\mu = \tan\lambda$), α is the tool rake angle, k is the shear strength of material, and A_s is the area of the shear plane.

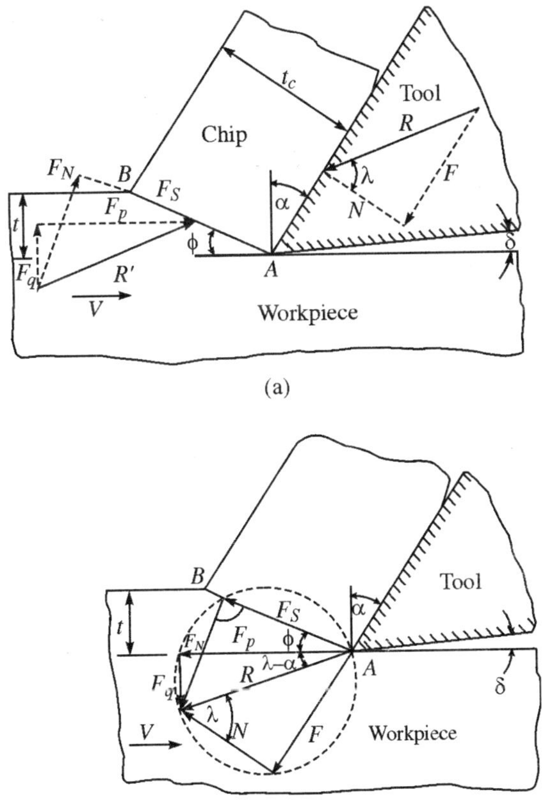

Fig. 4.1 Force diagram

If t is the underformed chip-thickness and b is the width of the workpiece, then

$$A_s = \frac{bt}{\sin\phi} \quad \text{...(4.3)}$$

From equations (4.1), (4.2) and (4.3),

$$R = \frac{kbt}{\sin\phi\cos(\phi+\lambda-\alpha)} \qquad ...(4.4)$$

Further,

$$F_p = R\cos\,(\lambda-\alpha) \qquad ...(4.5)$$

or

$$F_p = \frac{kbt\cos(\lambda-\alpha)}{\sin\phi\cos(\phi+\lambda-\alpha)} \qquad ...(4.6)$$

Similarly, thrust force

$$F_q = \frac{kbt\sin(\lambda-\alpha)}{\sin\phi\cos(\phi+\lambda-\alpha)} \qquad ...(4.7)$$

4.2.2 Shear Angle Relationship

Forces F_p and F_q can be calculated using equations (4.6) and (4.7) provided the yield shear stress (k), friction angle (λ) and shear angle (ϕ) are known. The variables b and t are known from the cutting conditions. If we idealize the workpiece material to be rigid, perfectly plastic, then k is known. Further, λ may be assumed to have the dry sliding friction value. To determine ϕ, Merchant (1944) applied the principle of minimum energy and suggested that during deformation, shear angle ϕ adjusts itself in such a way that the amount of energy consumed is minimized. This condition can be obtained by differentiating F_p with respect to ϕ and equating to zero. Thus,

$$\frac{dF_p}{d\phi} = \frac{kbt\cos(\lambda-\alpha)\cos(2\phi+\lambda-\alpha)}{\sin^2\phi\cos^2(\phi+\lambda-\alpha)} = 0 \qquad ...(4.8)$$

This gives

$$\phi = \frac{\pi}{4} - \frac{1}{2}(\lambda-\alpha) \qquad ...(4.9)$$

Using this shear angle relationship (equation 4.9), the theoretical values of cutting and thrust forces can be expressed in terms of ϕ as

$$F_p = \frac{kbt\cos(\lambda-\alpha)}{\sin\left[\frac{\pi}{4}-\frac{1}{2}(\lambda-\alpha)\right]\cos\left[\frac{\pi}{4}+\frac{1}{2}(\lambda-\alpha)\right]}$$

which on simplification gives

$$F_p = 2\,k\,bt\cot\phi \qquad ...(4.10)$$

Similarly,

$$F_q = \frac{kbt\sin(\lambda-\alpha)}{\sin\left[\frac{\pi}{4}-\frac{1}{2}(\lambda-\alpha)\right]\cos\left[\frac{\pi}{4}+\frac{1}{2}(\lambda-\alpha)\right]}$$

which finally gives

$$F_q = kbt(\cot^2\phi-1) \qquad ...(4.11)$$

In addition to assuming the minimum energy principle in deriving equations (4.9) to (4.11), it was also assumed that λ is a constant and is not influenced by the values of ϕ. Although there are good reasons to believe that this is not true but for simplicity it has been assumed that it is independent of ϕ.

The directly measurable quantities from among those shown in Fig. 4.1(b) are ϕ, F_p and F_q. While experimental methods of evaluating shear angle are discussed later in this chapter, the measurement of cutting forces is discussed in Chapter 6. From the experimental values of the above quantities, forces F, N, F_S and F_N can be evaluated using the following equations:

$$F = F_p \sin\alpha + F_q \cos\alpha \quad ...(4.12)$$

$$N = F_p \cos\alpha - F_q \sin\alpha \quad ...(4.13)$$

$$F_S = F_p \cos\phi - F_q \sin\phi \quad ...(4.14)$$

$$F_N = F_p \sin\phi + F_q \cos\phi \quad ...(4.15)$$

The values of forces thus obtained can now be used for determining the other quantities of interest such as coefficient of friction, shear-plane stresses, shear strain, shear strain rate, etc.

4.2.3 Determination of Coefficient of Friction

The forces acting on the rake face of the tool are of importance since they enable the coefficient of friction for the tool face to be determined. The coefficient of friction between two sliding surfaces is defined as the ratio of the force in the sliding direction to the force normal to the sliding interface. Thus, the coefficient of friction

$$\mu = \tan\lambda = \frac{F}{N}$$

$$= \frac{F_p \sin\alpha + F_q \cos\alpha}{F_p \cos\alpha - F_q \sin\alpha}$$

$$= \frac{F_p + F_q \tan\alpha}{F_p - F_q \tan\alpha} \quad ...(4.16)$$

Here it is implied that the forces F and N are uniformly distributed over the entrie chip-tool contact area. There is evidence to indicate that this is not true. The actual distribution of shear and normal stresses on the rake face of the tool is probably of the form shown in Fig. 4.2. From the tool tip to some point C, the normal stress is very high and the metal adheres to the tool face resulting in sticking friction conditions. From C to D where the chip curls, smaller normal stress values are obtained giving the usual sliding friction conditions.

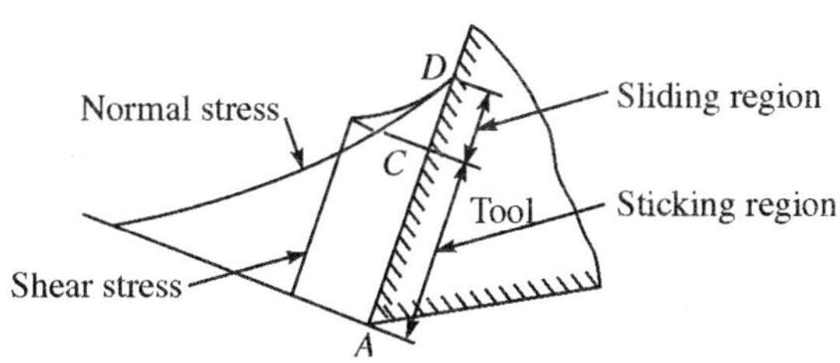

Fig. 4.2 Normal stress and shear stress on the rake face

The effect of normal force on the coefficient of friction is shown in Fig. 4.3. Under light friction conditions, that is, when the normal force is small, the friction force is proportional to the normal load and a constant coefficient of friction is obtained. The friction force and the coefficient of friction in this region are independent of the apparent area of contact between the sliding surfaces. In this region, only a few asperities come in contact as shown in Fig. 4.4(*a*). As a consequence of this, the real area of contact A_r is much smaller than the apparent area of contact A. For example, when two ground blocks of 1 sq. cm. cross-sectional area are placed in contact under a normal load of about 1 kg. the real area of contact is less than 0.01 percent of the apparent area of contact. As the load is increased the real area of contact increases initially due to elastic deformation of contacting asperities and then due to plastic deformation. Under conditions of high normal load, plastic deformation at sliding interfaces causes the real area of contact to approach the apparent area of contact. In the extreme case, the real area of contact will be equal to the apparent area of contact (Fig. 4.4*b*) and the friction force is independent of the applied normal load (Fig. 4.3).

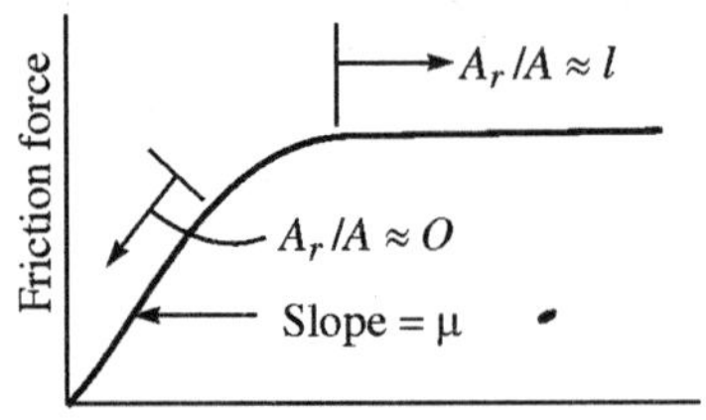

Fig. 4.3 Variation of friction force with normal force

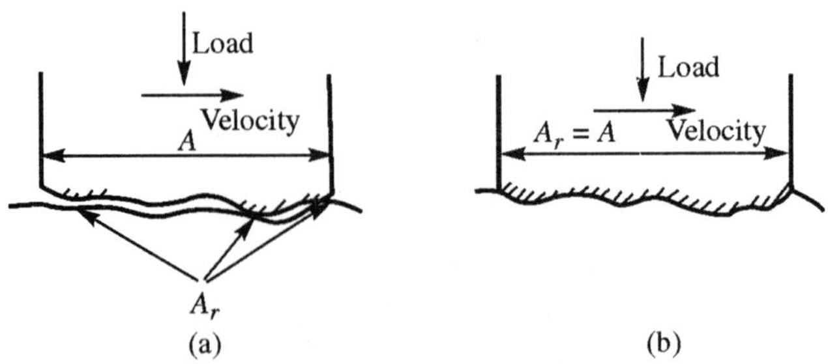

Fig. 4.4 Contact between mating surfaces in dry friction

4.2.4 Determination of Stress, Strain and Strain Rate

Using the experimental values of ϕ, F_p and F_q, the shear stress and the normal stress on the shear plane can be calculated from

$$\tau = \frac{F_s}{A_s} = \frac{(F_p \cos\phi - F_q \sin\phi)\sin\phi}{bt} \quad ...(4.17)$$

and

$$\sigma = \frac{F_N}{A_s} = \frac{(F_p \sin\phi + F_q \cos\phi)\sin\phi}{bt} \quad ...(4.18)$$

using equations (4.3), (4.14) and (4.15).

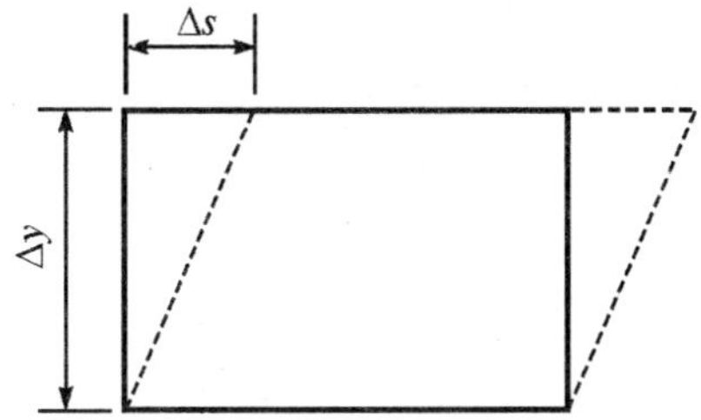

Fig. 4.5 Shear strain

The shear strain γ is defined as $\frac{\Delta s}{\Delta y}$ as shown in Fig. 4.5. Hence, in metal cutting (Fig. 4.6),

$$\gamma = \frac{\Delta s}{\Delta y} = \frac{AB'}{C'D'} = \frac{AD'}{C'D'} + \frac{D'B'}{C'D'}$$

or

$$\gamma = \tan(\phi - \alpha) + \cot\phi \qquad ...(4.19)$$

In deriving this expression, the deformation has been idealized as a process of block slip of preferred slip planes as shown in Fig. 4.6.

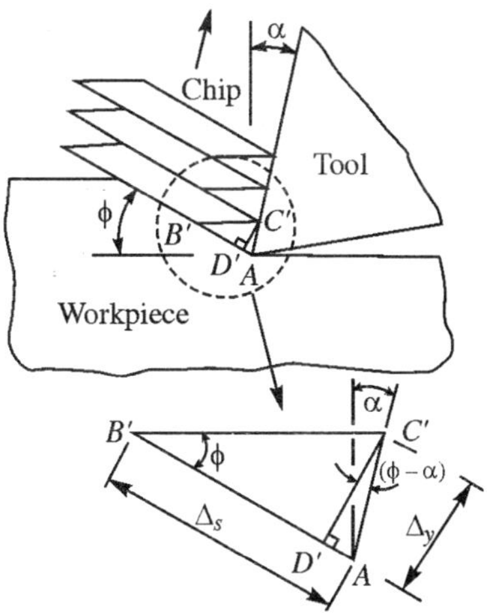

Fig. 4.6 Shear strain in metal cutting

The rate at which the shear strain $\dot{\gamma}$ is produced, is given by

$$\dot{\gamma} = \frac{\Delta s}{\Delta y} \cdot \frac{1}{\Delta t} \qquad ...(4.20)$$

where Δt is the time required for the metal to travel the distance Δs along the shear plane. Δy as before is the distance between two successive shear planes. $\dot{\gamma}$ can also be expressed in terms of shear velocity which can be obtained from the velocity diagram. Figure 4.7 gives the following velocity relationships:

$$V_c = \frac{\sin\phi}{\cos(\phi - \alpha)} \text{ V}; \; V_s = \frac{\cos\alpha}{\cos(\phi - \alpha)} \cdot V \qquad ...(4.21)$$

Hence,

$$\dot{\gamma} = \left(\frac{\Delta s}{\Delta t}\right) \cdot \frac{1}{\Delta y} = \frac{Vs}{\Delta y}$$

or

$$\dot{\gamma} = \frac{\cos\alpha}{\cos(\phi - \alpha)} \cdot \frac{V}{\Delta y} \qquad ...(4.22)$$

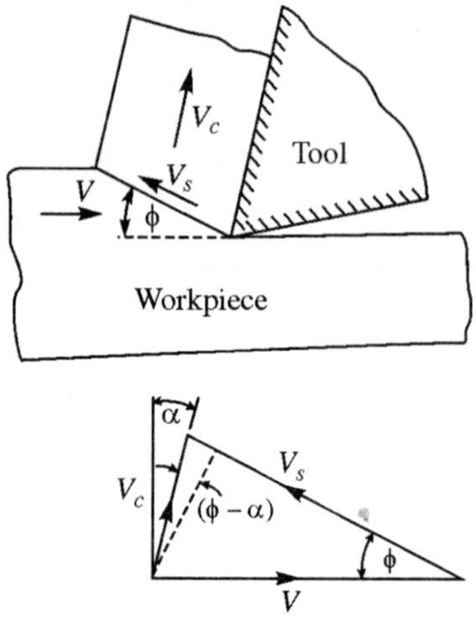

Fig. 4.7 Cutting velocities

From Fig. 4.6, it may appear that all atomic planes are active shear planes. This is not true. The strains are generally confined to planes that have a spacing of over 25×10^{-6} mm and a reasonable mean value of spacing between successive planes (Δy) would be around 25×10^{-4} mm.

The strain rate in machining is usually very high. For example, when $\alpha = 0°, \phi = 20°$ and $V = 0.5$ m/sec we find that $\dot{\gamma} \approx 2 \times 10^5$ sec^{-1}. Compare this with typical strain rate values of 10^{-3} sec^{-1} for tensile test and 10^3 sec^{-1} for impact test. The strain in metal cutting not only occurs at high rates but also takes place at reasonably high temperatures. A typical value of mean temperature in the shear plane is around 400 °C. These high values of strain rate and temperature will have significant effect on the material yield stress. In general, an increase in strain rate causes an increase in yield stress, while an increase in temperature results in decrease in yield stress. Thus, strain rate and temperature have opposite effect on yield stress, and under certain conditions may compensate each others effect. Again for simplicity we have neglected the effect of these parameters on the yield shear stress of workpiece material.

4.3 MEASUREMENT OF SHEAR ANGLE

Experimental methods of evaluating shear angle can be classified into two categories:

1. Direct method, and
2. Indirect method

The direct method involves photographing the shear plane on the side surface through a microscope using a high-speed camera. For better results, grid lines could be scribed on one of the side surfaces of the workpiece as shown in Fig. 4.8. Variation of this technique involves use of quick stop devices which freeze the deformation for examination. One such device is shown in Fig. 4.9. The workpiece *A* is held in the sliding cylinder *B* which slides inside the cylindrical groove provided in the body *C*. At the end of the cylindrical groove is a circular ring *D* which is attached to the body by means of shear pins *E*. During machining the workpiece is prevented from sliding by this ring. After the tool has moved some distance, the tongue *F* on the tool holder hits the end of the sliding cylinder and shears the pins. This separates the workpiece with partially formed chip for examination under the microscope.

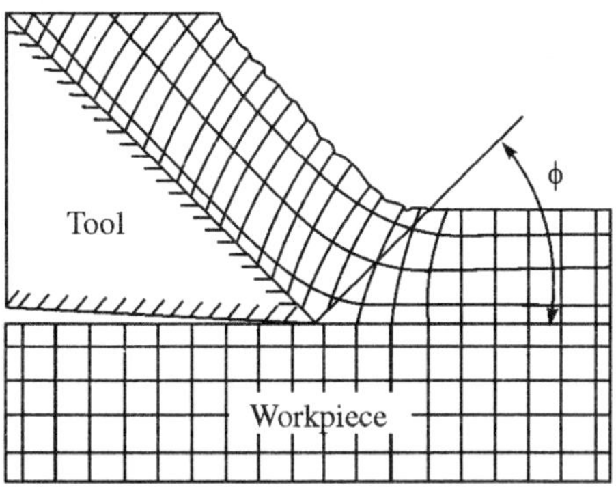

Fig. 4.8. Deformation of square grids during cutting

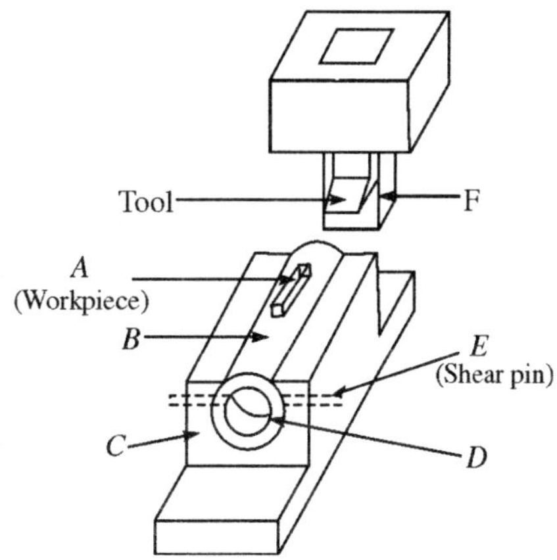

Fig. 4.9 Quick stop device

The indirect method of obtaining the shear angle is through the evaluation of chip-thickness ratio (r) which is defined as the ratio of underformed chip-thickness (t) to the chip thickness (t_c), *i.e.*,

$$r = \frac{t}{t_c} \qquad ...(4.23)$$

During metal cutting, which is a process of plastic deformation, there is practically no change in volume. Hence, from continuity of the workpiece and chip, we can write,

$$l\,t\,b = l_c\,t_c\,b_c \qquad ...(4.24)$$

where l is the length of cut, t is the thickness of undeformed chip and b is the width of cut. Subscript c refers to the corresponding measurements on the chip. In orthogonal cutting, there is no side spread, therefore, $b \approx b_c$. Hence,

$$\frac{l_c}{l} = \frac{t}{t_c} = r \qquad ...(4.25)$$

The relationship between shear angle ϕ and chip-thickness ratio r can be obtained from the geometry of cut (Fig. 4.10). It is evident that

$$AB = \frac{t}{\sin\phi} = \frac{t_c}{\cos(\phi - \alpha)} \qquad (4.26)$$

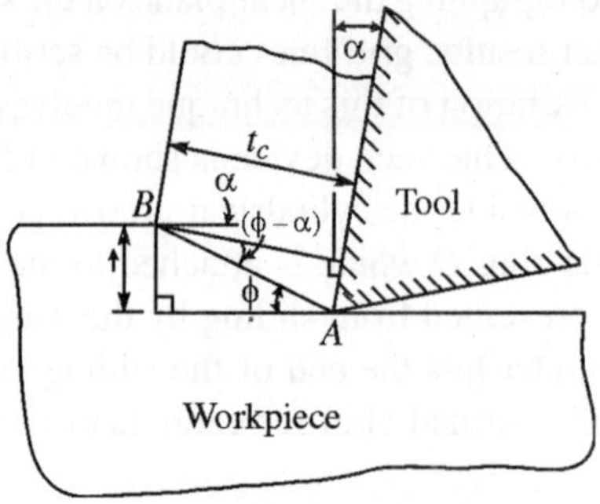

Fig. 4.10 The relationship between shear angle and chip thickness

or

$$r = \frac{\sin\phi}{\cos\phi\cos\alpha + \sin\phi\sin\alpha} \qquad ...(4.27)$$

Solving for ϕ, we get

$$\tan\phi = \frac{r\cos\alpha}{1 - r\sin\alpha} \qquad ...(4.28)$$

Thus, if the chip-thickness ratio is known, the shear angle can be evaluated.

While the value of r can be obtained by measuring the undeformed chip-thickness t and the thickness of the chip t_c, this method does not give precise results since the back of the chip is rough and it is difficult to obtain a representative value for deformed chip-thickness t_c. Evaluation of r from measurement of chip length is a better method since the chip-thickness ratio is also equal to the chip-length ratio (l_c/l).

The length of the workpiece in most cases are known. For cylindrical workpieces an axial slot can be cut on the test-piece which will cause a notch to be formed in the chip after each revolution (Fig. 4.11). The underformed chip-length can thus be calculated in terms of the workpiece diameter.

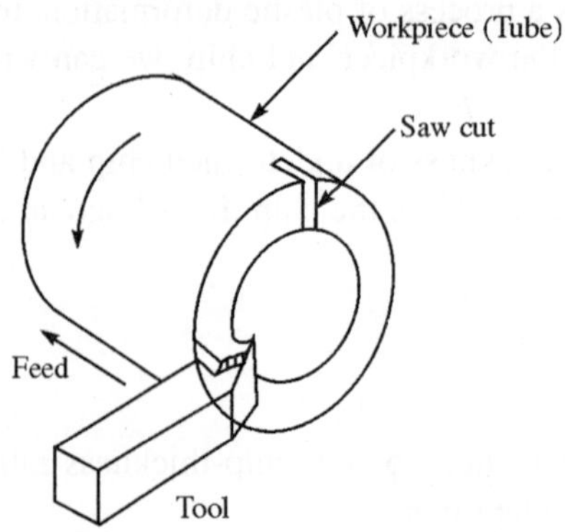

Fig. 4.11 Orthogonal cutting on lathe using tube

An alternative method is to weigh a known length of the chip and establish the length of cut. Weight of chip,

$$W_c = \rho l_c t_c b_c = \rho l t b \qquad ...(4.29)$$

where ρ is the density of workpiece material. Therefore,

$$l = \frac{W_c}{\rho b t} \qquad ...(4.30)$$

4.4 COMPARISON WITH EXPERIMENTS

The theory of metal cutting outlined above has been compared with experiments. A convenient way of making such a comparison is to plot the shear angle ϕ against $(\lambda - \alpha)$. On such a graph the relationship predicted by the theory will be a straight line. The experimental results for several metals are shown in Fig. 4.12, along with the predictions of the theory. The experimental values of ϕ have been obtained using the indirect method involving the evaluation of chip-length ratio (l_c / l) as described earlier. Tests have been carried out under orthogonal cutting conditions on tubes having an axial slot as shown in Fig. 4.11. Both experimental and theoretical results show linear relationship between ϕ and $(\lambda - \alpha)$. The experimental results show that there is no unique relationship of the kind predicted by the Merchant theory and the agreement between the theory and experiments is only on a qualitative sense.

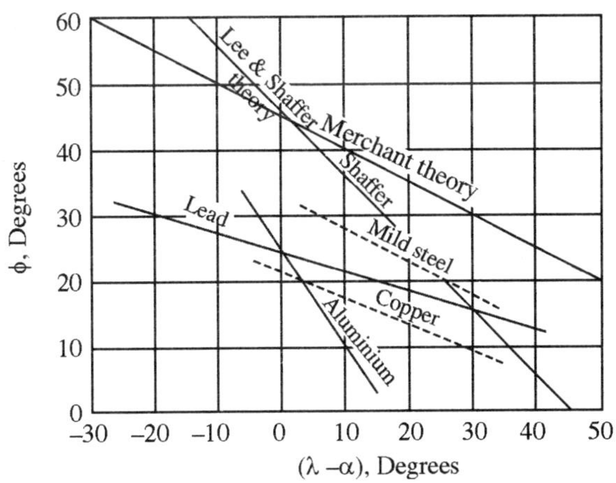

Fig. 4.12 Comparison with experimental results (After Pugh, 1958)

It is not surprising that the theoretical and experimental results do not show good agreement. In deriving the shear angle relationship, several simplifying assumptions have been made. The cutting edge has been assumed to be perfectly sharp which is a very rough approximation of the actual condition particularly when the undeformed chip-thickness is small. Observations have shown that even for ground tools the edge radius varies from 0.005 to 0.025 mm. Thus, the tool edge will tend to 'plough' through the work material rather than cut, resulting in additional forces on the clearance or flank face of the tool. It has been shown experimentally that these clearance faces forces (F_{cp} and F_{cq}) are quite significant (Fig. 4.13).

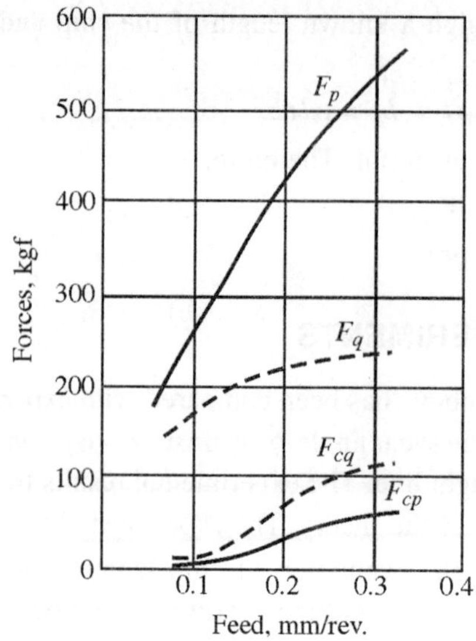

Fig. 4.13 Variation of clearance face forces with feed (After Spaans, 1967)

Experimental results also show that the assumption of a shear plane is a rough approximation in the low-cutting speed range. The shape of the plastic (shear) zone obtained experimentally is shown in Fig. 4.14. Also, the coefficient of friction on the rake face of the tool is not constant and the assumption of mean coefficient of friction is again a rough approximation of the actual conditions. This aspect has been discussed earlier. Lastly, the effects of strain hardening, strain rate and temperature on yield stress of work material have not been considered in the theoretical analysis.

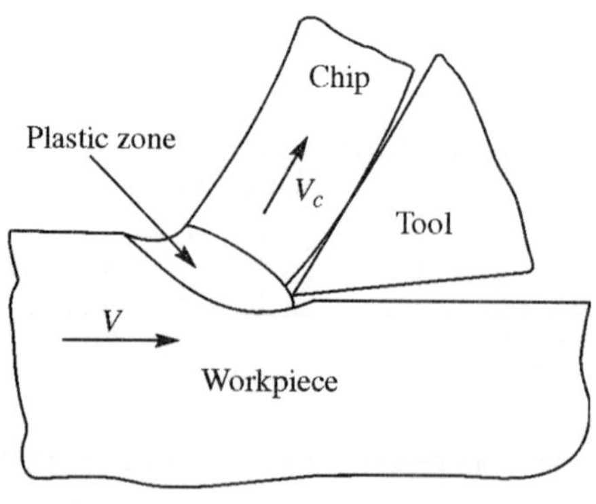

Fig. 4.14 Shear zone in metal cutting

Inspite of many simplifying assumptions, the theory predicts values for forces etc. which are useful for most practical purposes. Taking appropriate factors of safety, the theoretical results can be used by the designer of machine tools in deciding about the size and shape of machine components, power requirements, etc. It is also useful in understanding the effect of cutting conditions such as rake angle, undeformed chip-thickness, etc. on the output parameters such as force, stress, strain, etc. which affect the surface integrity of workpiece, and useful life of tool. Several other investigators have proposed ingenious and intricate models for evaluating the shear angle, some of them have been quite successful, but the main advantage fo Merchants' analysis presented here is its simplicity.

4.5 OTHER ANALYSES

The theory of machining outlined earlier is essentially based on the equilibrium of forces associated with the cutting process. Since the presentation of this model by Merchant (1944), several authors have applied the theory of plasticity to analyse the problem of orthogonal machining. One simple analysis by Lee and Shaffer (1951) who assumed a uniform plastic zone in the chip and applied the slip-line field theory to obtain the shear angle relationship for orthogonal cutting. Oxley (1961) also used the slip-time field concept to develop his analysis.

4.5.1 Slip-line Field Method

The slip-line field analysis, a graphical approach, depends upon the determination of the plastic flow pattern in the deforming material. Plastic flow, as discussed earlier, occurs predominantly by slip on planes of closely-packed atoms and in the direction of the line of atoms which lie closest to the line of maximum shear stress. In a real material, it is reasonable to assume that there will be sufficient number of favourably oriented planes for the slip direction to coincide with the direction of maximum shear stress. In an ideal material, *i.e.*, a structureless, homogeneous and isotropic material, the slip will always occur precisely in the direction of maximum shear stress. Thus, once the direction of maximum shear stress is known, the direction of plastic flow in an ideal material is known.

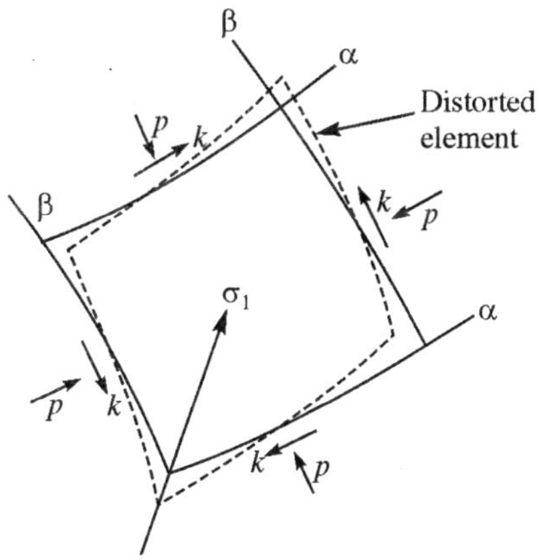

Fig. 4.15 Curvilinear element of a slip-line field

In a deforming material the planes of maximum shear stress from an orthogonal curvilinear network as shown in Fig. 4.15. These orthogonal network of characteristics or the lines of maximum shear stress have come to be known as the slip-line field and the directions of maximum shear stress as the slip lines. In a rigid plastic material, each element is subjected to the maximum shear stress k and a mean normal stress (hydrostatic pressure) p. The direction of principal stresses σ_1 and σ_3 are at 45° to the slip lines. The third principal stress σ_2 acts normal to the plane of the figure. The general problem of three-dimensional deformation is still intractable, but it is possible to analyse plane strain (two-dimensional) problems using the slip-line field approach. In a plane strain situation, the strain in the direction of the third principal stress is zero. Further, there can be no normal stress at a free surface and components of the shear stresses along the slip lines when resolved parallel to the surface must be equal. The slip lines, therefore, meet a free surface at 45° (Fig. 4.16).

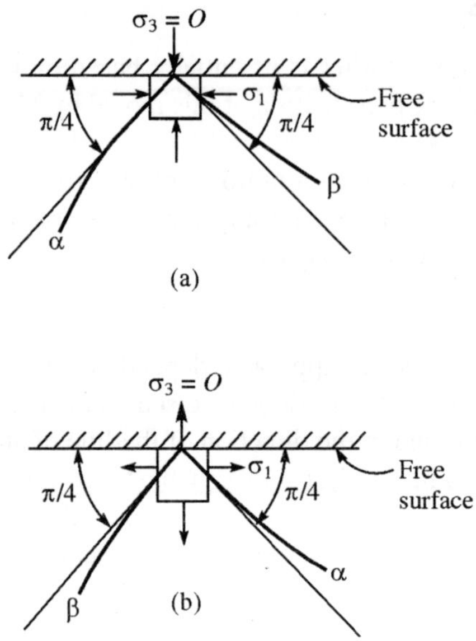

Fig. 4.16 Slip lines at a free surface

The two directions of maximum shear stresses or the slip lines are usually identified as α and β-lines. The usual convention for identifying is that when α and β-lines form a right-handed co-ordinate system of axes, then the direction of the algebraically greatest principal stress (say σ_1, as shown in Fig. 4.17) lies in the first and third quadrants. The counter clock-wise rotation ψ of the α-line from the chosen x-direction is taken as positive. The Hencky equations,

$$p + 2k\psi = \text{constant (along an } \alpha \text{-line)} \qquad ...(4.31)$$

and

$$p - 2k\psi = \text{constant (along a } \beta \text{-line)} \qquad ...(4.32)$$

enable the hydrostatic pressure p at any point in the deforming region to be evaluated from the curvature of the slip lines. The values of the constants are obtained from known boundary conditions.

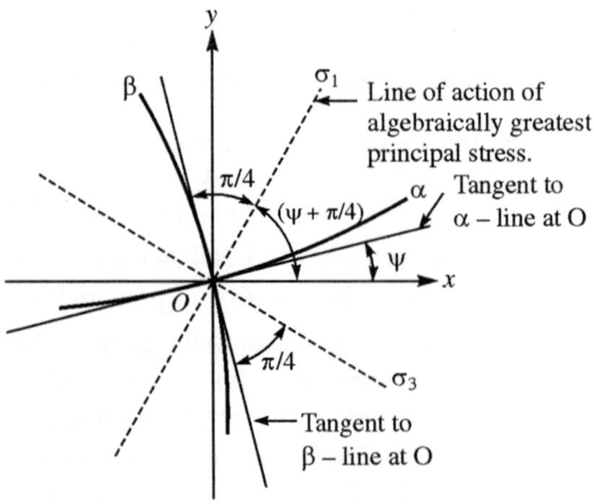

Fig. 4.17 The stresses on a curvilinear element

The major feature of the slip-line field theory concerns the manner in which solutions are arrived at. The usual sense in which solutions are obtained does not apply here. Instead of deriving the solution, the solution is first proposed and then it is shown to be kinematically admissible and that it satisfies all boundary conditions. Obviously, the solution obtained will not be a unique one and many times other criterion, such as minimum work, have to be applied to arrive at the best solution.

4.5.2 Lee and Shaffer Analysis

Lee and Shaffer (1951) were perhaps the first to analyse the orthogonal cutting problem using a slip-line field and this model appears to be the simplest of all such solutions. Their derivation of the shear angle relationship is described here to illustrate the application of slip-line field technique to machining problem.

The slip-line field proposed by Lee and Shaffer for orthogonal cutting is shown in Fig. 4.18(a). Their solution is based on the following assumptions:

(*i*) Deformation takes place under plane strain conditions.
(*ii*) Deformation occurs in a thin zone (shear plane).
(*iii*) The workpiece material is rigid, perfectly plastic and its behaviour is independent of temperature and strain rate.
(*iv*) Inertia effects are negligible (steady-state deformation).
(*v*) Tool tip is sharp (zero nose radius).
(*vi*) Continuous chip is formed.
(*vii*) Stresses are uniform on the rake face (constant coefficient of friction).

It is clear that they have also employed the shear plane model with the shear plane *AB* making an angle ϕ with the direction of cutting velocity. The cutting forces are transmitted through the triangular plastic zone *ABC* where no deformation occcurs but the entire material is in the plastic state with maximum shear stress *k* everywhere. Clearly, the shear plane *AB* is a slip line since maximum shear stress must occur along the shear plane. The other slip line must be perpendicular at this line. Since no forces are transmitted to the chip after it has crossed the line *BC,* no stress can occur across this boundary. Thus, *BC* may be regarded as a free surface and the slip lines must meet this surface at 45°.

Mohr circle construction is a convenient means of relating stresses on any plane to the principal stresses. In the two-dimensional case, the stress is represented by any point on the circumference of the circle. Figure 4.18(b) shows the Mohr circle for the stress zone. Since plane *BC* is stress free, the Mohr circle must pass through the origin. Further, the Mohr circle should be a compressive circle since the stresses involved are compressive. The origin is marked *b* in accordance with the letters on the physical plane and points *a*, *c*, *d* and *f* arc displaced from *b* by 90° (twice the angle of physical plane). Face *e* is inclined to face *d* at an angle $(\phi - \alpha)$, therefore, in the stress plane the angle subtended by arc *ae* at the centre is $2(\phi - \alpha)$.

Assuming uniform shear stress τ and normal stress σ on the rake face, the friction angle,

$$\lambda = \tan^{-1}\left(\frac{\tau}{\sigma}\right) = \tan^{-1}\left(\frac{es}{sb}\right) = L\, ebs \qquad \ldots(4.33)$$

and

$$L\, eos = 2\,\lambda \qquad ...(4.34)$$

The shear angle relationship can now be obtained by adding angles *eos* and *eod*. Thus,

$$2\lambda + 2(\phi - \lambda) = \frac{\pi}{2} \qquad ...(4.35)$$

or

$$\phi = \frac{\pi}{4} - (\lambda - \alpha) \qquad ...(4.36)$$

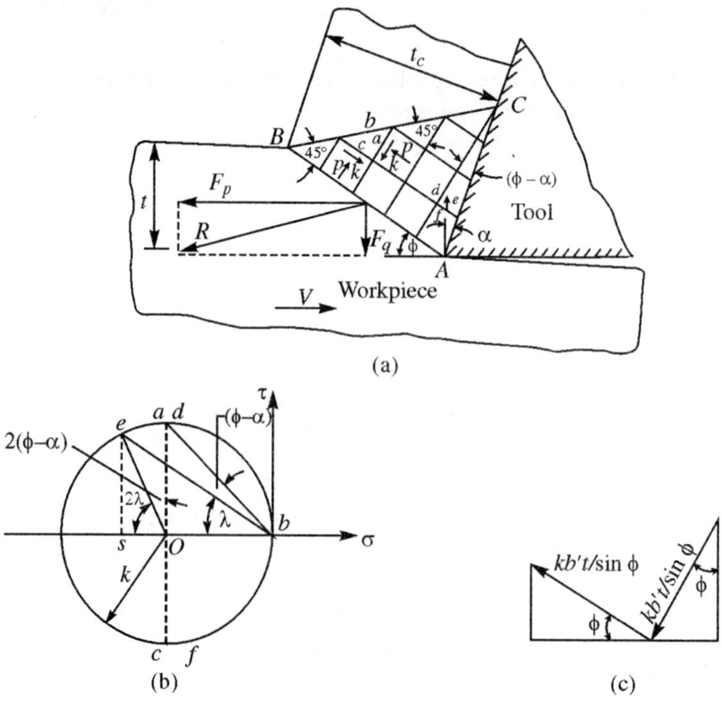

Fig. 4.18 Lee and Shaffer's solution; (*a*) Physical plane, (*b*) Mohr's circle for the stress zone and (*c*) Diagram for estimating cutting and thrust forces.

The forces can be found by evaluating the shear and normal stresses acting on *AB*. Since the slip lines are straight ($\psi = 0$ in equations 4.31 and 4.32), the hydrostatic stress at *A* and *B* are equal, *i.e.*, $p_A = p_B$. The hydrostatic pressure p_B can be obtained by examining the 'free surface' *BC* where the principal stress normal to the surface is zero and Mohr circle for stress gives $p_B = k$. Therefore, both shear and normal stresses on *AB* are equal to *k* and both tangential and normal forces are equal to $k\,bt/\sin\phi$ as shown in Fig. 4.18(c). Here b' is the width of the workpiece and *t* is the undeformed chip-thickness. The cutting force F_p in the direction of cutting velocity *V* and the thrust force F_q perpendicular to the work surface can now be written as

$$F_P = \frac{kb't}{\sin\phi}\cos\phi + \frac{kb't}{\sin\phi}\sin\phi$$

$$= kb't(\cot\phi + 1) \qquad ...(4.37)$$

and $$F_q = \frac{kb't}{\sin\phi}\cos\phi - \frac{kb't}{\sin\phi}\sin\phi$$

$$= kb't(\cot\phi - 1) \qquad ...(4.38)$$

The shear angle relationship (equation 4.36) is not valid for all the value of α and λ. For example, when $\alpha = 0$ and $\lambda = \pi/4, \phi$ would be zero. Lee and Shaffer have, however, suggested that such conditions of low rake angle and high friction would lead to built-up-edge formation and the solution proposed is not valid. For such conditions they have proposed an alternate solution.

For comparison with experiments, Lee and Shaffer's shear angle values have also been plotted in Fig. 4.12. This analysis also shows a linear relationship between ϕ and $(\lambda - \alpha)$ but the agreement with experiements is no better than those predicted by Merchant's shear angle relationship. Merchant analysis is, however, much simpler and in general terms apply for quite a large range of cutting conditions.

4.5.3 Oxley Analysis

Palmer and Oxley (1959) studied the deformation pattern during slow-speed machining by recording the paths of individual grains on the side surface of the workpiece using a cine film camera. The shape of the plastic zone obtained is shown in Fig. 4.14. In their subsequent analyses (Oxley *et al.* 1961, 1963), they assumed the deformation to occur along a thin shear zone. Available experimental evidence also indicates that in the practical cutting speed range the deformation is confined to a narrow zone, which can be approximated to be a thin shear plane.

Oxley (1961) assumed the narrow deformation zone to be bounded by straight and parallel slip lines inclined at an angle ϕ to the direction of cutting as shown in Fig. 4.19. Slip line CD is assumed to turn upward at D to meet the free surface at 45° to satisfy the stress boundary condition. Since the deformation is assumed to occur along a thin shear plane, point B on the shear plane (slip line) AB can be assumed to meet the free surface at 45°. Point B is asumed to be at negligible distance from the free surface.

Oxley used the generalized form of Hencky equations,

$$p + 2k\psi + \int \frac{\partial k}{\partial s_\beta} ds_\alpha - 2\int \psi \frac{\partial k}{\partial s_\alpha} ds_\alpha = \text{constant (along an } \alpha\text{-line)} \qquad ...(4.39)$$

$$p - 2k\psi + \int \frac{\partial k}{\partial s_\alpha} ds_\beta + 2\int \psi \frac{\partial k}{\partial s_\beta} ds_\beta = \text{constant (along a } \beta\text{-line)} \qquad ...(4.40)$$

to evaluate the stress conditions along the shear plane AB. Neglecting the work hardening term along the slip lines, the last term in the above equations becomes zero and the stress equations reduce to

$$p + 2k\psi + \int \frac{\partial k}{\partial s_\beta} ds_\alpha = \text{constant (along an } \alpha\text{-line)} \qquad ...(4.41)$$

$$p + 2k\psi + \int \frac{\partial k}{\partial s_\alpha} ds_\beta = \text{constant (along a } \beta\text{-line)} \qquad ...(4.42)$$

In these equations, ds_α and ds_β are elemental lengths along α-and β-slip lines, respectively. For a rigid, perfectly plastic material, the above equations reduce to the form shown in equations (4.31) and (4.32).

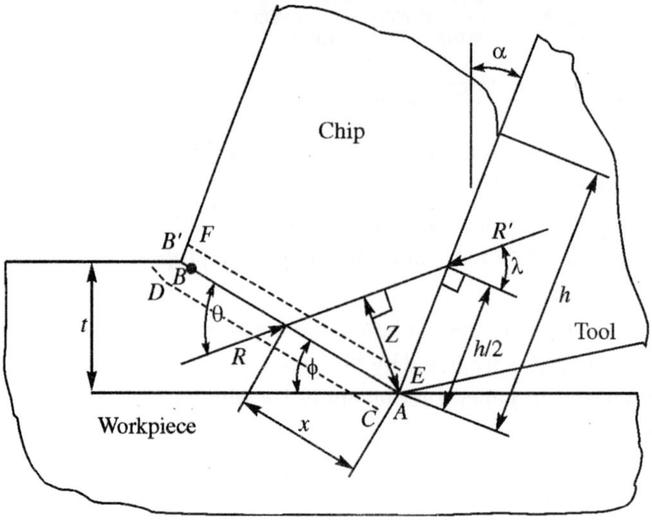

Fig. 4.19 Oxley's model

The hydrostatic stress at B (p_B) can now be found using equation (4.41) or (4.42) after identifying the slip line along the shear plane as α or β line. Since the principal stress at the free surface is zero and the other principal stress is compressive, the slip line along the shear plane is identified as β-line as shown in Fig. 4.16(a). Using equation (4.42),

$$p_B - 2k\psi_B = p_{B'} - 2k\psi_{B'} \qquad \text{...(4.43)}$$

neglecting the work hardening term because B is at negligible distance from B' at the free surface.

Taking $\psi_{B'} = 0$, *i.e.*, along the co-ordinate axis, $\psi_{B'} = \left(\frac{\pi}{4} - \phi\right)$ and $p_{B'} = k$ from Mohr's Circle (Fig. 4.20),

$$p_B = k\left[1 + 2\left(\frac{\pi}{4} - \phi\right)\right] \qquad \text{...(4.44)}$$

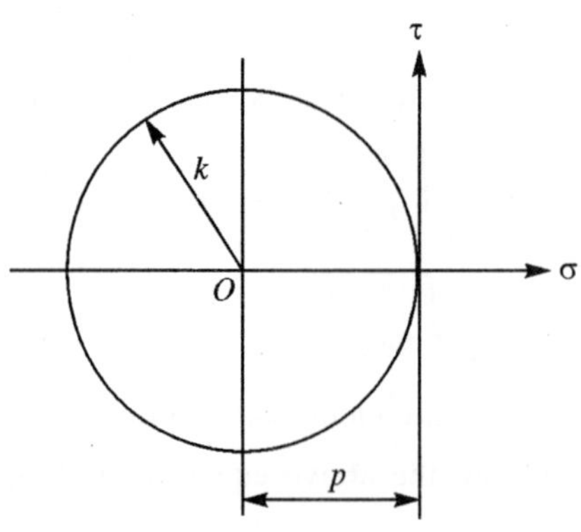

Fig. 4.20 Mohr's circle at B'

The stresses on slip line AB can be related to the stresses on the rake face of the tool. From Mohr's stress circle at A (Fig. 4.21),

$$[p_A + k\sin 2(\phi-\alpha)]\tan\lambda = k\cos 2(\phi-\alpha) \quad ...(4.45)$$

or

$$p_A = k\left[\frac{\cos 2(\phi-\alpha)}{\tan\lambda} - \sin 2(\phi-\alpha)\right] \quad ...(4.46)$$

where the friction angle λ is assumed to be constant along the rake face and ϕ and α are the shear and rake angles, respectively.

Assuming linear hydrostatic stress distribution on AB with $p_B > p_A$, the direction of resultant force θ on AB can be evaluated from

$$\tan\theta = \frac{\frac{1}{2}(p_A + p_B)}{k} \quad ...(4.47)$$

Solution of equation (4.44), (4.46) and (4.47) gives

$$\theta = \tan^{-1}\left[\frac{1}{2} + \frac{\pi}{4} - \phi + \frac{\cos 2(\phi-\alpha)}{2\tan\lambda} - \frac{\sin 2(\phi-\alpha)}{2}\right] \quad ...(4.48)$$

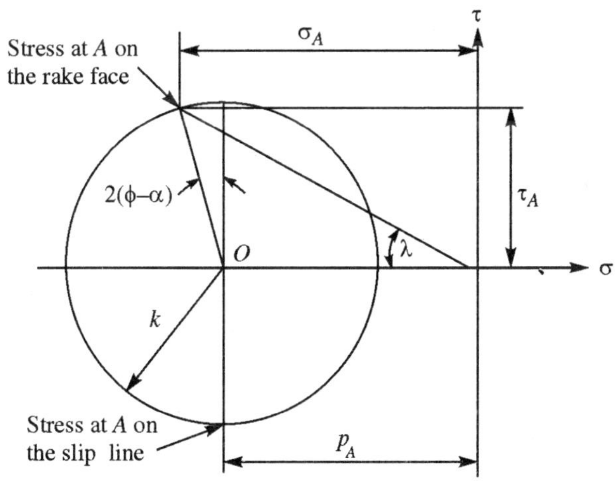

Fig. 4.21 Mohr's circle at A

For equilibrium of chip, the resultant forces R on the shear plane and the resultant force R' on the rake face must be equal, opposite and co-linear. Thus, in Fig. 4.19 $R = R'$.

Since the normal and tangential stresses on the rake face are assumed to be uniformly distributed (constant λ), the angle of R' with the cutting direction is $(\lambda - \alpha)$. Hence, from Fig. 4.19

$$\theta = \phi + \lambda - \alpha \quad ...(4.49)$$

Equations (4.48) and (4.49) can now be solved to evaluate the shear angle ϕ for any given values of α and λ.

The validity of the assumed stress distribution can be checked by comparing the moment of the resultant force evaluated independently from the stress on the shear plane and the rake face. The moment of stresses acting on AB about B (Fig. 4.19) is $\frac{1}{2}(p_A + p_B)x$.

Distance x can be evaluated from the hydrostatic stress distribution on shear plane AB. From Fig. 4.22,

$$x_1 = \frac{1}{2}AB = \frac{t}{2\sin\phi} \quad ...(4.50)$$

$$x_2 = \frac{2}{3}AB = \frac{2t}{3\sin\phi} \quad ...(4.51)$$

Therefore,

$$p_A \cdot \frac{t}{\sin\phi}x_1 + \frac{1}{2}(p_A + p_B)\frac{t}{\sin\phi}x_2 = \frac{1}{2}(p_B - p_A)\frac{t}{\sin\phi}x \quad ...(4.52)$$

or

$$x = \frac{\left(\frac{p_A}{6} + \frac{p_B}{3}\right)\frac{t}{\sin\phi}}{\left(\frac{p_A + p_B}{2}\right)} \quad ...(4.53)$$

The moment arm Z can now be evaluated as

$$Z = x \sin\theta \quad ...(4.54)$$

or

$$Z = \frac{2t\left(\frac{p_A}{6} + \frac{p_B}{3}\right)\sin\theta}{(p_A + p_B)\sin\theta} \quad ...(4.55)$$

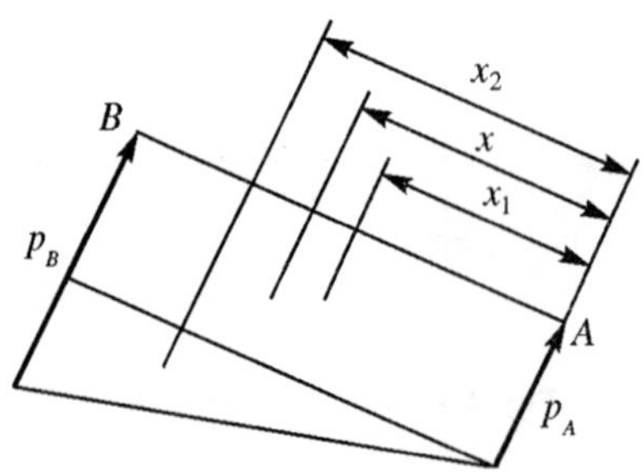

Fig. 4.22 Hydrostatic stress distribution

Since the normal stress (σ) and tangential stress (τ) are assumed to be uniform along the rake face, the resultant force R' on the rake face will act at the mid-point of the tool-chip contact interface, *i.e.*, at $\frac{1}{2}h$ (Fig. 4.19). The normal force (N) on the rake face is given by

$$N = \sigma bh \quad ...(4.56)$$

where b is the width of cut. Also,

$$N = R'\cos\lambda = R\cos\lambda \quad ...(4.57)$$

or

$$N = \left(\frac{kbt}{\sin\phi\cos\theta}\right)\cos\lambda \qquad \text{...(4.58)}$$

From equations (4.56) and (4.58),

$$h = \frac{kt\cos\lambda}{\sigma\sin\phi\cos\theta} \qquad \text{..(4.59)}$$

This assumed interface stress distribution gives, the moment arm Z' for the resultant force R' as

$$Z' = \frac{h}{2}\cos\lambda \qquad \text{...(4.60)}$$

or

$$Z' = \frac{kt\cos^2\lambda}{2\sigma\sin\phi\cos\theta} \qquad \text{...(4.61)}$$

Oxley compared the values of Z and Z' calculated from equations (4.55) and (4.61) for a range of values of λ determined experimentally and the corresponding calculated values of ϕ. His results (Fig. 4.23) show that Z and Z' values are fairly close.

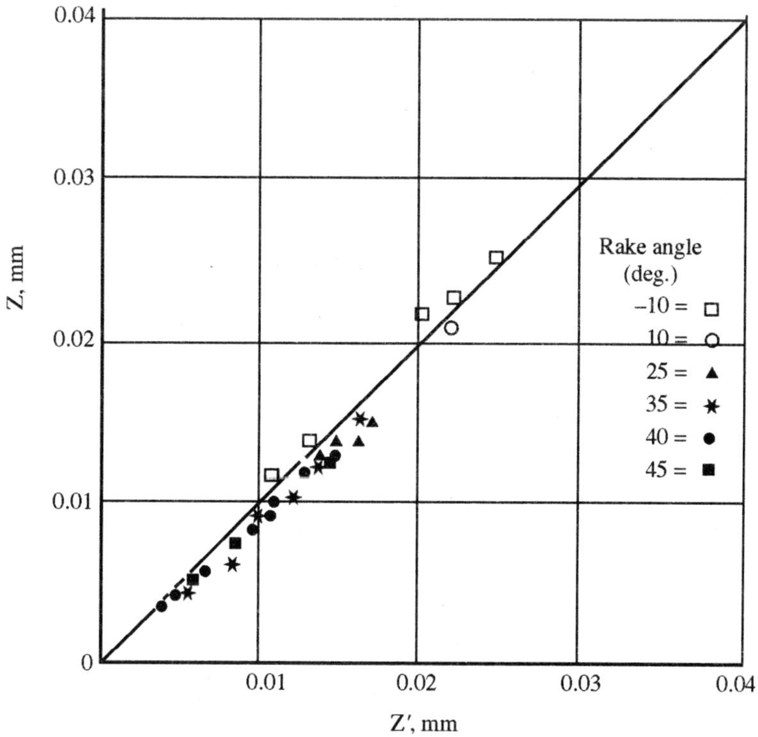

Fig. 4.23 Comparision of moment arms Z and Z' calculated from shrear zone and tool-chip interface stresses (After Oxley, 1961)

4.5.4 Extension of Oxley Analysis

Oxley and Hatton (1963) extended the earlier analysis by introducing non-uniform stress distribution on the rake face. The actual rake-face stress distribution (Fig. 4.2) was simplified as shown in Fig. 4.24. The average friction angle λ can now be defined by

$$\tan\lambda = \frac{F}{N} \quad \text{...(4.62)}$$

where

$$F = \tau_A h' b + \frac{1}{2}\tau_A (h - h')b$$

$$= \frac{1}{2}\tau_A (h + h')b \quad \text{...(4.63)}$$

and

$$N = \frac{1}{2}\sigma_A hb \quad \text{...(4.64)}$$

These equations give

$$\tan\lambda = \frac{\tau_A}{\sigma_A}\left(1 + \frac{h'}{h}\right) \quad \text{...(4.65)}$$

or

$$\frac{\tau_A}{\sigma_A} = \frac{\tan\lambda}{(1 + h'/h)} \quad \text{...(4.66)}$$

From Mohr circle at A (Fig. 4.21)

$$\tan\lambda_A = \frac{\tau_A}{\sigma_A} = \frac{\tan\lambda}{\left(1 + \dfrac{h'}{h}\right)} \quad \text{...(4.67)}$$

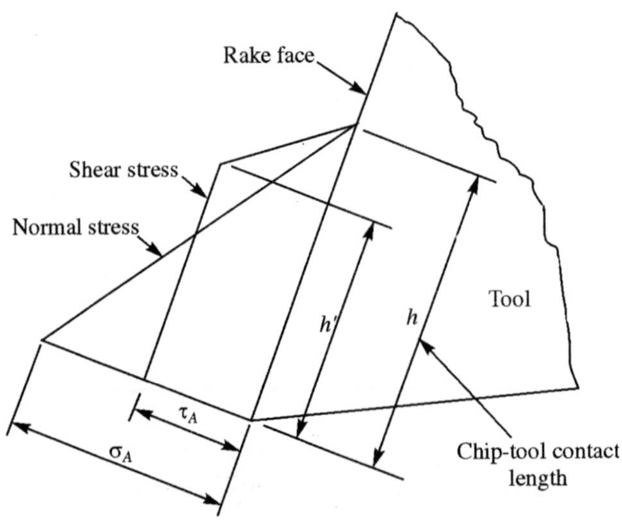

Fig. 4.24 Stress distribution on rake face.

p_B, p_A and $\tan\theta$ are again given by equations (4.44), (4.46) and (4.47), respectively. Solution of equations (4.44), (4.46), (4.47) and (4.67) gives

$$\theta = \tan^{-1}\left[\frac{1}{2}+\frac{\pi}{4}-\phi+\left(1+\frac{h'}{h}\right)\frac{\cos 2(\phi-\alpha)}{2\tan\lambda} - \frac{\sin^2(\phi-\alpha)}{2\tan\lambda}-\frac{\sin 2(\phi-\alpha)}{2}\right] \quad ...(4.68)$$

The equation can now be solved alongwith equation (4.49) to obtain values of ϕ for given values of α, λ and h'/h. In this case also the plot of Z and Z' gave close agreement.

When values of α (10° to 45°), tan λ (0.8 to 2.0) and h'/h (0.5 to 1) are plotted against $(\lambda - \alpha)$, a linear relationship is obtained (Fig. 4.25). The relationship can be approximated as

$$\phi = 50° - 0.8(\lambda - \alpha) \quad ...(4.69)$$

This relationship appears to give values of ϕ which are somewhat closer to the measured values of shear angle than those predicted by equation (4.9) and (4.36).

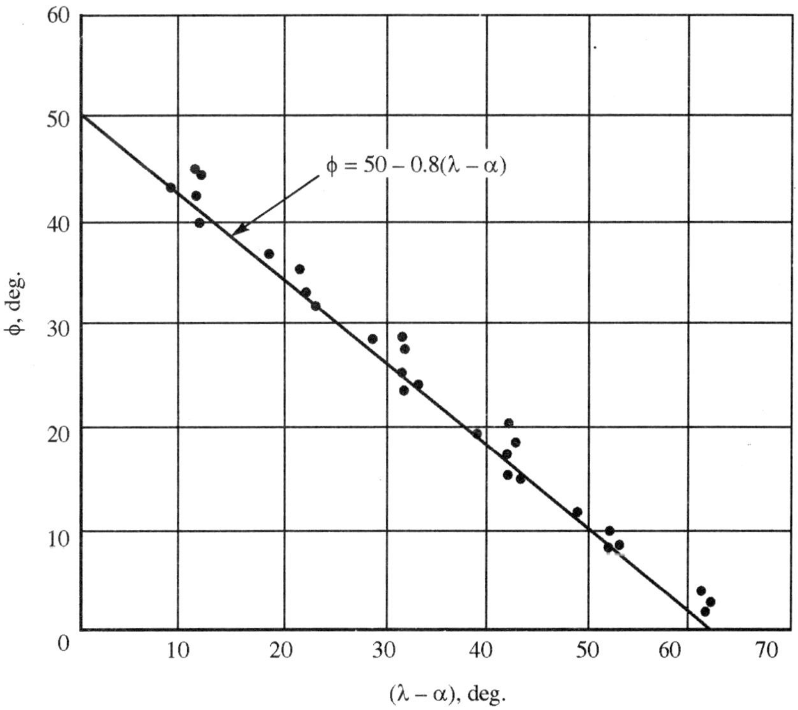

Fig. 4.25 Solution of equations (4.65) and (4.68) for values of α = 10 to 40 deg., tan λ = 0.8 to 2.0 and h'/h = 0.5 to 1 (After Armarego and Brown, 1969).

CHAPTER 5

Mechanics of Oblique Cutting

5.1 INTRODUCTION

In the oblique cutting process, the tool cutting edge is inclined at an angle other than 90° to the direction of cutting as shown in Fig. 2.9. Although oblique cutting is the most general type of cutting, the mechanics involved in the process have not been investigated to the same extent as the simpler orthogonal cutting process. Merchant (1944) was perhaps the first to propose the basic model for oblique cutting based on the principle of minimum energy. Later Stabler (1955), Brown and Armarego (1964), Spaans (1967, 1970), Morcos (1972) and others analysed the process to predict chip-flow angle, shear angle and forces.

The analysis of orthogonal cutting has indicated that rake angle is a fundamental variable influencing the forces and shear angle. Rake angle in oblique cutting can be measured in three different planes as discussed in section 3.1. The rake angles corresponding to these three planes are normal rake angle (α_n), velocity rake angle (α_v) and effective rake angle (α_e) as shown in Fig. 3.1. These three angles are obviously related and from the geometry, this relationship can be obtained as

$$\tan\alpha_v = \frac{\tan\alpha_n}{\cos i} \qquad ...(5.1)$$

and

$$\sin\alpha_e = \sin i \sin\eta_c + \cos i \cos\eta_c \sin\alpha_n \qquad ...(5.2)$$

where i and η_c are the inclination and chip-flow angles.

Experiments show that the force in the direction of cutting velocity vector (cutting or power force) remains constant with constant normal rake angle, but it varies with α_v and α_e. Thus, α_n appears to be the angle corresponding to the rake angle in orthogonal cutting.

Considering the shear plane to extend from the cutting edge to the work surface, two definitions of shear angle have been suggested. The angle from the shear plane to the plane containing the newly formed work surface measured in a plane normal to the cutting edge is called the *normal shear angle* (ϕ_n). When the angle of shear plane is measured in the plane containing the cutting velocity (V_w) and the chip velocity (V_c) vectors, the angle is called the *effective shear angle* (ϕ_e). Following the orthogonal cutting approach (equation 4.28), the normal shear angle (ϕ_n) and effective shear angle (ϕ_e) can be evaluated from

$$\tan\phi_n = \frac{r_t \cos\alpha_n}{1 - r_t \sin\alpha_n} \qquad ...(5.3)$$

and

$$\tan\phi_e = \frac{V_c / V_w \cos\alpha_e}{1 - V_c / V_w \sin\alpha_e}$$

$$= \frac{r_t r_b \cos\alpha_e}{1 - r_t r_b \sin\alpha_e} \quad ...(5.4)$$

Here the chip-thickness ratio $r_t = t / t_c$ and chip-width ratio $r_b = b / b_c$ and the incompressibility condition gives

$$r_t r_b = \frac{t}{t_c}\frac{b}{b_c} = \frac{V_c}{V_w} \quad ...(5.5)$$

For orthogonal cutting, both equations (5.3) and (5.4) reduce to equation (4.28) since $i = 0$, $\alpha_n = \alpha_e = \alpha, r_b = 1$ and $\phi_n = \phi_e = \phi$.

5.2 FORCE RELATIONSHIPS

In oblique cutting, it is convenient to resolve the resultant force R into three components. The axes that are generally considered are P along the cutting velocity vector, Q perpendicular to the work surface and R perpendicular to P and Q as shown in Fig. 5.1(*a*). In Fig. 5.2, these force components are shown as F_P, F_Q and F_R. For convenience the resultant force R can also be resolved into two components R' in a plane perpendicular to the cutting edge and R'' along the cutting edge. Force component R' can be further resolved as F_s' and F_N' or F' and N (Fig. 5.3). Therefore,

$$R = \sqrt{F_P^2 + F_Q^2 + F_R^2}$$

$$= \sqrt{(R')^2 + (R'')^2}$$

$$= \sqrt{(F_P')^2 + (F_Q')^2 + (F_R')^2} \quad ...(5.6)$$

since $R'' = F_R'$.

From Fig. 5.2, the force component F_p is given by

$$F_P = F_P' \cos i + F_R' \sin i$$

$$= F_S'\left[\frac{\cos(\lambda_n - \alpha_n)\cos i}{\cos(\phi_n + \lambda_n - \alpha_n)} + F_S \sin\eta_s \sin i\right]$$

$$= F_S\left[\frac{\cos(\lambda_n - \alpha_n)\cos i \cos\eta_s}{\cos(\phi_n + \lambda_n - \alpha_n)} + \sin\eta_s \sin i\right] \quad ...(5.7)$$

Here F_S is the shear force in the shear plane inclined at an angle η_s to the normal to the cutting edge in the shear plane. The normal friction angle λ_n (corresponding to α_n and ϕ_n) is given by

$$\tan\lambda_n = \frac{F'}{N} = \frac{F}{N}\cos\eta_c = \tan\lambda \cos\eta_c \quad ...(5.8)$$

where λ is the friction angle and

$$\tan\lambda = \frac{F}{N} \quad ...(5.9)$$

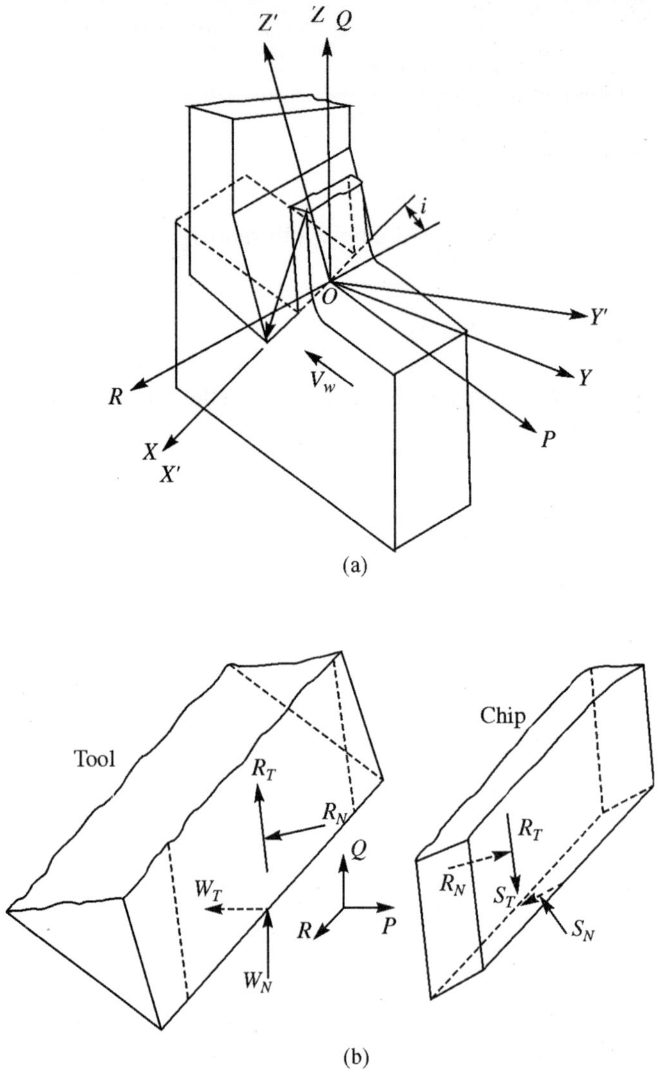

Fig. 5.1 (*a*) Oblique cutting with a single-edged tool and (*b*) equilibrium of chip

The direction of shear force η_s is obtained from

$$\tan\eta_s = \frac{F'_R}{F'_S} = \frac{F'\tan\eta_c}{R'\cos(\phi_n+\lambda_n-\alpha_n)}$$

$$= \frac{\tan\eta_c \sin\lambda_n}{\cos(\phi_n+\lambda_n-\alpha_n)} \quad \text{...(5.10)}$$

The shear force on the shear plane is

$$F_S = kA_s = k\frac{bt}{\sin\phi_n \cos i} \quad \text{...(5.11)}$$

where k is the shear stress on the shear plane and A_s is the area of shear plane. Using equation (5.10) and (5.11), equation (5.7) reduces to

$$F_P = \frac{kbt}{\sin\phi_n}\left[\frac{\cos(\lambda_n - \alpha_n) + \tan i \tan\eta_c \tan\lambda_n}{\sqrt{\cos^2(\phi_n + \lambda_n - \alpha_n) + \tan^2\eta_c \sin^2\lambda_n}}\right] \quad ...(5.12)$$

The force component F_Q can be evaluated from

$$F_Q = F'_Q = R'\sin(\lambda_n - \alpha_n)$$

$$= \frac{F'_S \sin(\lambda_n - \alpha_n)}{\cos(\phi_n + \lambda_n - \alpha_n)}$$

$$= \frac{F_S \cos\eta_s \sin(\lambda_n - \alpha_n)}{\cos(\phi_n + \lambda_n - \alpha_n)} \quad ...(5.13)$$

Substituting for η_s and F_S from equation (5.10) and (5.11), equation (5.13) reduces to

$$F_Q = \frac{kbt}{\sin\phi_n \cos i}\frac{\sin(\lambda_n - \alpha_n)}{\sqrt{\cos^2(\phi_n + \lambda_n - \alpha_n) + \tan^2\eta_c \sin^2\lambda_n}} \quad ...(5.14)$$

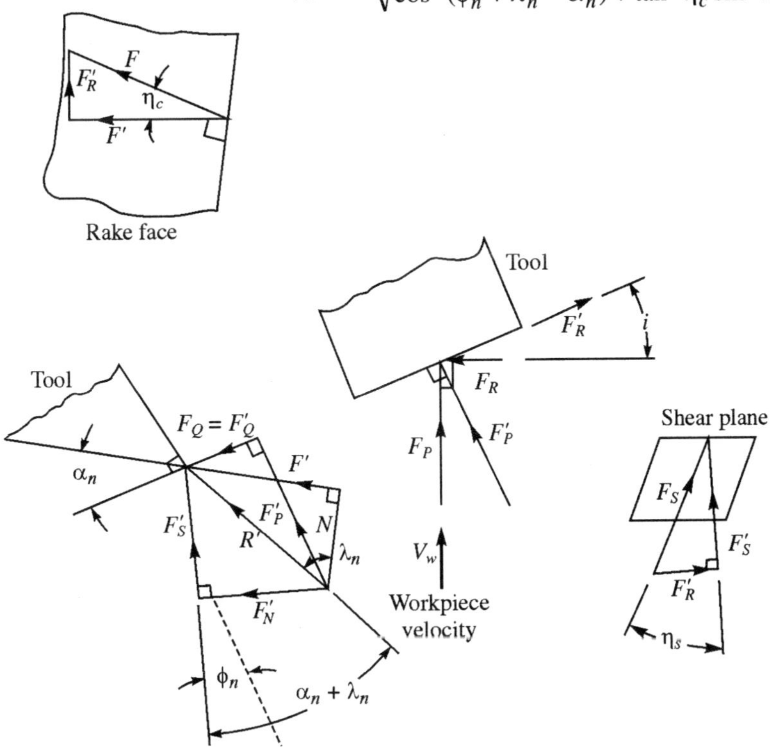

Fig. 5.2 Force components in oblique cutting

From Fig. 5.2, the force component F_R is

$$F_R = F'_P \sin i - F'_R \cos i$$

$$= F_S\left[\frac{\cos(\lambda_n - \alpha_n)\cos\eta_s \sin i}{\cos(\phi_n + \lambda_n - \alpha_n)} - \sin\eta_s \cos i\right] \quad ...(5.15)$$

Again substituting for η_s and F_S, equation (5.15) gives

$$F_R = \frac{kbt}{\sin\phi_n}\left[\frac{\cos(\lambda_n - \alpha_n)\tan i - \tan\eta_c \sin\lambda_n}{\sqrt{\cos^2(\phi_n + \lambda_n - \alpha_n) + \tan^2\eta_c \sin^2\lambda_n}}\right] \quad ...(5.16)$$

These equations for force component were first derived by Merchant (1944). Forces F_P, F_Q and F_R can now be evaluated if shear angle ϕ_n, chip-flow angle η_c and friction angle λ are known.

5.3 VELOCITY RELATIONSHIPS

Referring to Fig. 5.3, the three velocity components are the cutting velocity or the workpiece velocity V_w, the chip velocity V_c on the rake face and shear velocity V_s on the shear plane. These three velocity vectors are in the same plane and are related as

$$V_c = V_w \frac{\sin\phi_e}{\cos(\phi_e - \alpha_e)} \quad ...(5.17)$$

and

$$V_s = V_w \frac{\cos\alpha_e}{\cos(\phi_e - \alpha_e)} \quad ...(5.18)$$

The relationship between velocity components can also be obtained by resolving them into three directions; (a) parallel to the cutting edge in the plane of the workpiece, (b) normal to the cutting edge in the plane of the workpiece and (c) normal to the work surface. These are shown as X, Y and Z axes in Fig. 5.1.

Resolving along X-direction,

$$V_w \sin i - V_c \sin\eta_c - V_s \sin\eta_s = 0 \quad ...(5.19)$$

Resolving along Y-direction,

$$V_w \cos i - V_c \cos\eta_c \sin\alpha_n - V_s \cos\eta_s \cos\phi_n = 0 \quad ...(5.20)$$

Resolving along Z-direction,

$$-V_c \cos\eta_c \cos\alpha_n + V_s \cos\eta_s \sin\phi_n = 0 \quad ...(5.21)$$

Solution of these equations gives

$$V_c = V_w \frac{\sin\phi_n \cos i}{\cos(\phi_n - \alpha_n)\cos\eta_c} \quad ...(5.22)$$

and

$$V_s = V_w \frac{\cos i \cos\alpha_n}{\cos\eta_s \cos(\phi_n - \alpha_n)} \quad ...(5.23)$$

Substituting the above two equations into equation (5.19) gives

$$\tan\eta_s = \frac{\tan i \cos(\phi_n - \alpha_n) - \tan\eta_c \sin\phi_n}{\cos\alpha_n} \quad ...(5.24)$$

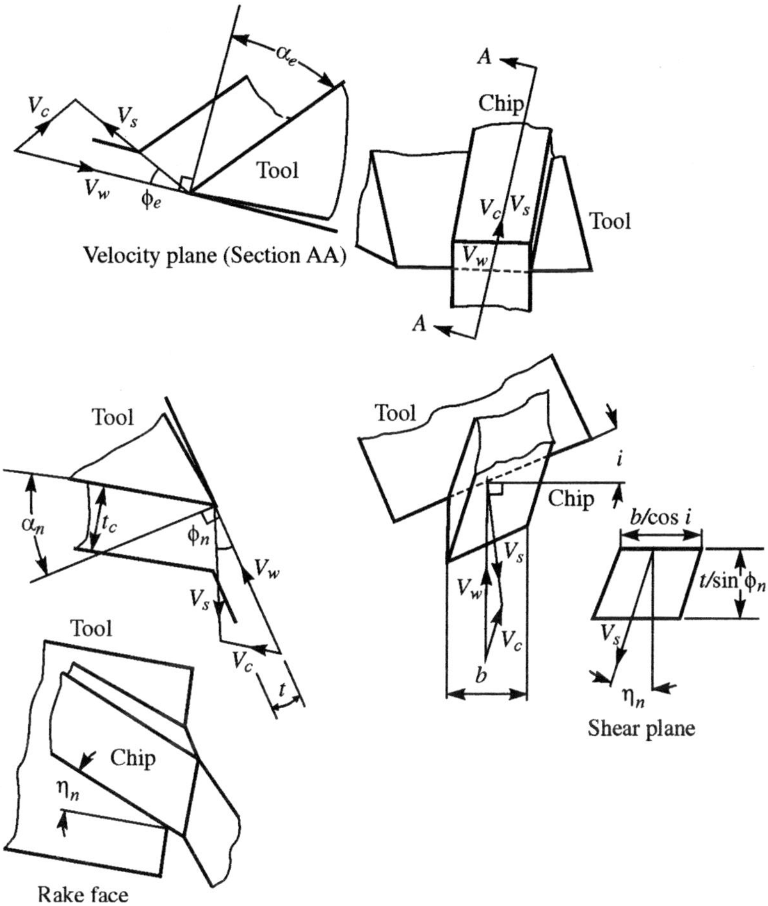

Fig. 5.3 Velocity components in oblique cutting

5.4 SHEAR ANGLE RELATIONSHIP

The shear angle relationship in oblique cutting can now be obtained by equating equations (5.10) and (5.24). This gives

$$\frac{\tan\eta_c \sin\lambda_n}{\cos(\phi_n + \lambda_n - \alpha_n)} = \frac{\tan i \cos(\phi_n - \alpha_n) - \tan\eta_c \sin\alpha_n}{\tan\eta_c - \sin\alpha_n \tan i}$$

which on simplification gives the shear angle relationship,

$$\tan(\phi_n + \lambda_n) = \frac{\cos\alpha_n \tan i}{\tan\eta_c - \sin\alpha_n \tan i} \quad ...(5.25)$$

The normal shear angle ϕ_n can be evaluated from the above equation provided λ or λ_n and η_c are known.

5.5 COEFFICIENT OF FRICTION

Force components F_P, F_Q and F_R can be measured with the help of a suitably designed three-component force dynamometer. From these measured values, the friction and normal force component F and N can be easily evaluated.

From Fig. 5.2,

$$F = \sqrt{[(F_P \cos i + F_R \sin i)\sin\alpha_n + F_Q \cos\alpha_n]^2 + (F_P \sin i - F_R \cos i)^2} \quad ...(5.26)$$

and

$$N = (F_P \cos i + F_R \sin i)\cos\alpha_n - F_Q \sin\alpha_n \quad ...(5.27)$$

The coefficient of friction μ can now be evaluated as

$$\mu = \tan\lambda = \frac{F}{N} \quad ...(5.28)$$

5.6 STRESS, STRAIN AND STRAIN RATE

Using the measured values of force components F_P, F_Q, F_R and ϕ_n, the shear force and the normal force on the shear plane F_S and F_N can be obtained from Fig. 5.2 as

$$F_S = \sqrt{[(F_P \cos i + F_R \sin i)\cos\phi_n - F_Q \sin\phi_n]^2 + (F_P \sin i - F_R \cos i)^2} \quad ...(5.29)$$

and

$$F_N = (F_P \cos i + F_R \sin i)\sin\phi_n + F_Q \cos\phi_n \quad ...(5.30)$$

The area of shear plane,

$$A_s = \frac{bt}{\cos i \sin\phi_n} \quad ...(5.31)$$

The shear stress τ and the normal stress σ can now be obtained as

$$\tau = \frac{F_S}{A_s} \quad ...(5.32)$$

and

$$\sigma = \frac{F_N}{A_s} \quad ...(5.33)$$

Considering block-type slip process during chip formation as in the case of orthogonal cutting (Fig. 4.6), the strain and strain rate can be evaluated. Referring to Fig. 5.4, the shear strain γ_n normal to the cutting edge is $\Delta s/\Delta y$ and the strain γ parallel to the shear velocity V_s is given by

$$\gamma = \frac{\Delta s}{\Delta y \cos\eta_s}$$

$$= \frac{\cot\phi_n + \tan(\phi_n - \alpha_n)}{\cos\eta_s} \quad ...(5.34)$$

The shear strain rate

$$\dot{\gamma} = \frac{\Delta s}{\Delta y \cos\eta_s}\frac{1}{\Delta t}$$

$$= \frac{V_s}{\Delta y} \qquad \text{...(5.35)}$$

where Δt is the time required for the metal to move the distance Δs on the shear plane.

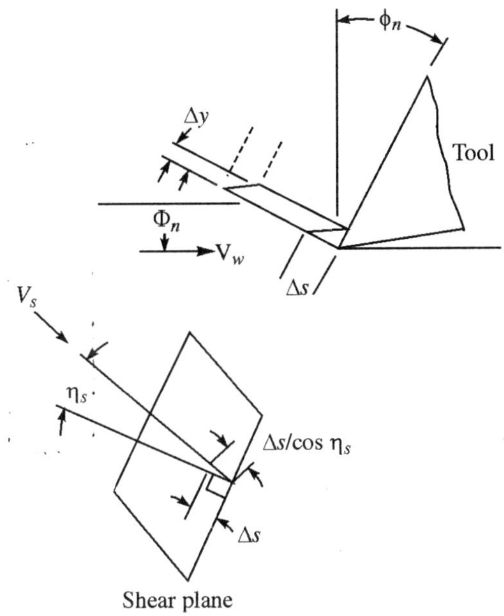

Fig. 5.4 Evaluation of shear strain

5.7 MEASUREMENT OF CHIP-FLOW DIRECTION

A number of methods have been used to measure the direction of chip flow during oblique cutting. These methods can be broadly classified into two categories — indirect measurements and direct measurements.

Assuming the component of the resultant force in the plane of rake face to be equal and opposite to the friction force, the chip-flow direction can be obtained. The resultant force on the rake face has two components — a component normal to the rake face and a friction component parallel to the rake face. The resultant force can be evaluated from force measurements using a suitable three-component dynamometer.

The chip-flow angle can also be evaluated from measurements of chip width using the relationship

$$\frac{b}{b_c} = \frac{\cos i}{\cos \eta_c} \qquad \text{...(5.36)}$$

obtained from Fig. 5.5. This chip-width method of evaluating the chip-flow angle, however, involves considerable difficulties in measuring b_c because of side flow in the chip.

Several direct methods have been used for measuring η_c. One method involves the introduction of a small amount of very fine abrasive powder under the chip and measuring the angle of scratches produced. Another method involves staining the rake face by Prussian blue and recording the marks produced. These methods are, however, not satisfactory since the chip-tool contact is over a very narrow region and fluctuations in chip movement give false indication of chip-flow direction.

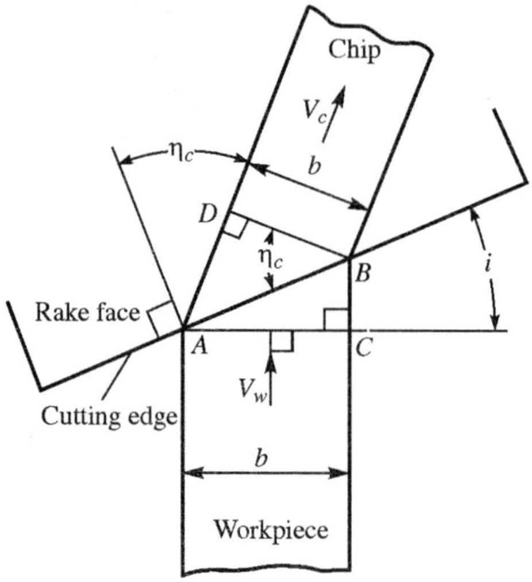

Fig. 5.5 Evaluation of chip-flow angle from measurement of chip width

The direct measurement using a microscope is not practical hence use of quick-stop device or photography appears to provide the best answer. Use of quick-stop device such as one shown in Fig. 4.9 freezes the deformation for examination. By using a proper light source and fast film, the photography method can be successfully used for recording the chip-flow direction even at higher cutting speeds. This method records instantaneous and not average values. A typical photograph of chip sliding over the rake face is shown in Fig. 5.6.

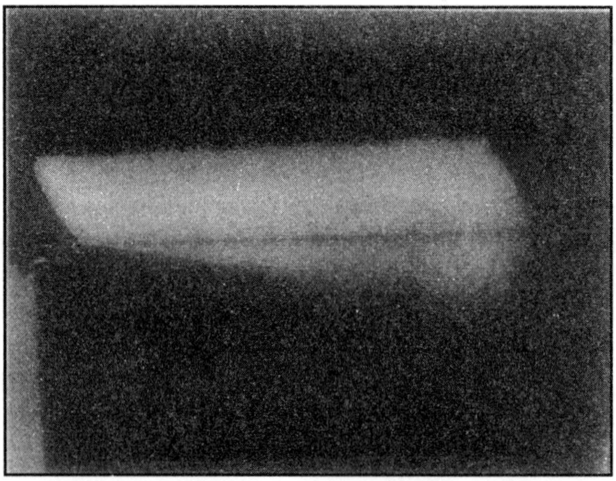

Fig. 5.6 Photograph of a partially formed chip

Stabler (1955) proposed a relationship based on his experimental results that

$$\eta_c = i \qquad ...(5.37)$$

Later he modified this as $\eta_c = 0.9i$ and suggested that this correlates better with experimental data.

Russell and Brown (1966) suggested that chip-flow angle is influenced by the normal rake angle and proposed the relationship

$$\tan\eta_c = \tan i \cos\alpha_n \quad ...(5.38)$$

Morcos (1972) indicated that the chip-flow angle is not only a function of the cutting geometry but is also influenced by the cutting conditions characterized by the shear angle. His expression for chip flow was

$$\tan\eta_c = \frac{\cos(\phi_n - \alpha_n)}{\sin\phi_n}\tan i \quad ...(5.39)$$

Experimental results showing the variation of η_c with i and α_n (Figs. 5.7 and 5.8) indicate that the relationship between η_c and i is neither linear nor independent of α_n. Table 5.1 indicates that the flow rule proposed by Morcos is closest to the experimental values but it cannot predict η_c from experimental conditions since it requires the knowledge of shear angle ϕ_n. Stabler flow rule, on the other hand, appears to be very handy and may be sufficiently accurate for most practical purposes.

Table 5.1 Comparison of chip-flow angle (After Lal, 1982)

			η_c			
No.	i	α_n	*(a)*	*(b)*	*(c)*	*(d)*
Material: Brass						
1	10	0	15.27	10	10	17.4
2	20	0	27.43	20	20	32.9
3	30	0	41.18	30	30	45.6
4	40	0	46.08	40	40	53.7
5	10	10	13.92	10	9.85	16.2
6	20	10	27.54	20	19.72	31.8
7	30	10	38.34	30	29.62	43.3
8	40	10	46.51	40	39.57	52.3
9	20	20	28.71	20	18.88	31.1
10	10	30	12.99	10	8.68	14.2
11	20	30	24.66	20	17.49	28.8
12	30	30	35.10	30	26.56	39.2
13	40	30	44.58	40	36.01	48.4
14	20	40	22.69	20	15.58	26.5
Material: Mild steel						
1	10	10	15.97	10	9.85	18.20
2	20	10	26.94	20	19.72	31.40
3	30	10	39.40	30	29.62	39.99
4	40	10	48.35	40	39.57	47.97
5	10	30	11.60	10	8.68	12.02
6	20	30	20.80	20	17.49	23.20
7	30	30	32.04	30	26.56	33.10
8	40	30	40.60	40	36.01	40.96

(a) Experimental, (b) Stabler, 1951 (Equation 5.37), (c) Russell and Brown, 1966 (Equation 5.38) and (d) Morcos, 1972 (Equation 5.39)

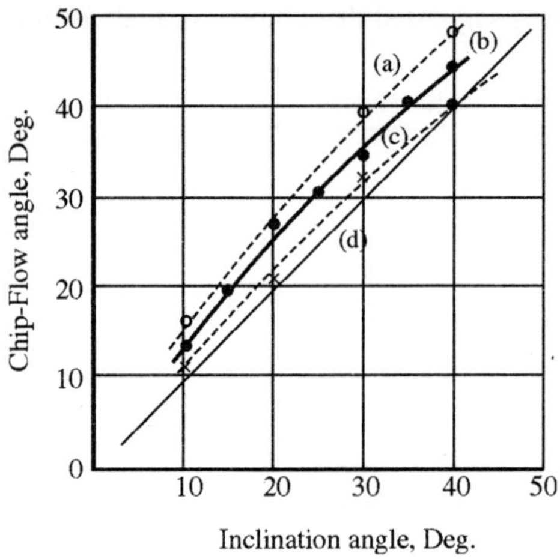

Fig. 5.7 Variation of η_c and i. (*a*) Mild steel, $\alpha_n = 10$, (*b*) Brass, $\alpha_n = 30$, (*c*) Mild steel, $\alpha_n = 30$ and (d) $\eta_c = i$ (After Lal, 1982)

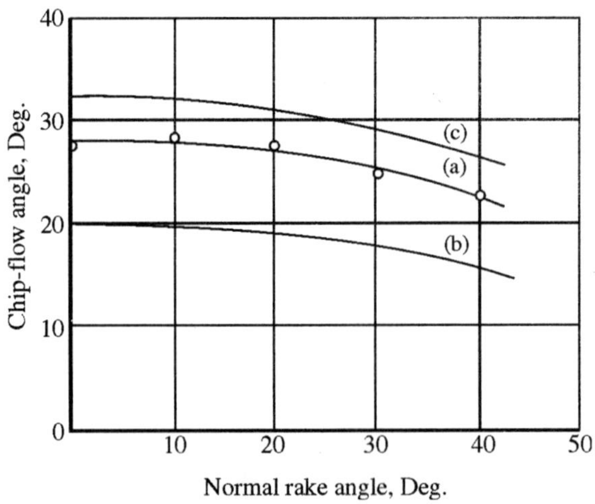

Fig. 5.8 Variation of η_c with α_n. (*a*) Brass, $i = 20$, (*b*) equation (5.38), (*c*) equation (5.39), (After Lal, 1982)

5.8 MEASUREMENT OF SHEAR ANGLE

One of the most popular method of measuring the shear angle is through the evaluation of chip-thickness ratio r_t.

The incompressibility condition gives

$$lbt = l_c b_c t_c \quad ...(5.40)$$

as in equation (4.24).

The chip-thickness ratio,

$$r_t = \frac{t}{t_c} = \frac{b_c}{b}\frac{l_c}{l} = r_c \frac{\cos\eta_c}{\cos i} \quad ...(5.41)$$

Here $r_c \left(= \frac{l_c}{l} \right)$ is the chip-length ratio. Substituting for r_t in equation (5.3), the normal shear angle can now be obtained.

If Stabler's chip-flow rule, $\eta_c = i$ (equation 5.37) is substituted in equation (5.25), it reduces to

$$\tan(\phi_n + \lambda_n) = \frac{\cos\alpha_n}{1 - \sin\alpha_n} \quad ...(5.42)$$

for which

$$\phi_n = \frac{\pi}{4} - \lambda_n + \frac{\alpha_n}{2} \quad ...(5.43)$$

is a solution. This equation is similar to the shear angle relationships given by equations (4.9) and (4.36) for orthogonal cutting.

5.9 PARASITIC FORCES

The cutting edge is never perfectly sharp and the tool-tip roundness causes the material to be pushed between the flank face of the tool and the work surface. This force neither takes part in the chip-rake face contact nor contributes to the shearing process. Similarly, the friction forces due to the wear of flank face and those due to presence of built-up-edge do not contribute to the shearing process. These have often been called '*parasitic*' forces and under certain cutting condition, their magnitude could be significant.

Spaans (1967, 1970) suggested a method of determining these forces from measurements of cutting forces, chip-flow angle and chip-length ratio. His method considers the equilibrium of forces on the tool and the chip and are expressed in terms of six linear equations, three for the tool and three for the chip. The forces involved are indicated in Fig. 5.1 (*b*) and can be expressed in terms of measurable parameters.

The force, resulting from stress caused by the contact of the chip with the rake face, has a component perpendicular to the rake face R_N and a component tangential to the rake plane in the chip flow direction R_T. There is no component perpendicular to these two components.

The force, resulting from the stress near the cutting edge which differs from the chip-tool contact stress, has a component W_T in the direction of the cutting velocity and a component W_N perpendicular to both the cutting velocity and the cutting edge. There is no component perpendicular to these two components.

The force resulting from the shear stresses on the shear plane S_T acts in the direction of shear velocity. The force resulting from normal stresses on the shear plane S_N is perpendicular to the shear plane.

Resolving the forces along, say P, Q and R directions, six linear equations of force equilibrium are obtained in terms of α_n, i, ϕ_n and η_c. When cutting conditions and the geometry of the tool are known, it is sufficient to measure the chip-length ratio (r_c) and the chip-flow angle (η_c) simultaneously to obtain the forces acting on the flank face under those conditions.

Figure 5.9 shows the rake and flank face forces evaluated through the solution of six equations of equilibrium of forces as suggested by Spaans. Experimentally measured values of forces, chip-length values and chip-flow angles have been used for obtainig the solution. A negative force component in this figure is merely an indication of the direction. It is clear that the flank face forces are significant.

Also, at higher depths of cut the flank forces increase because of increased flank surface area in contact with the workpiece.

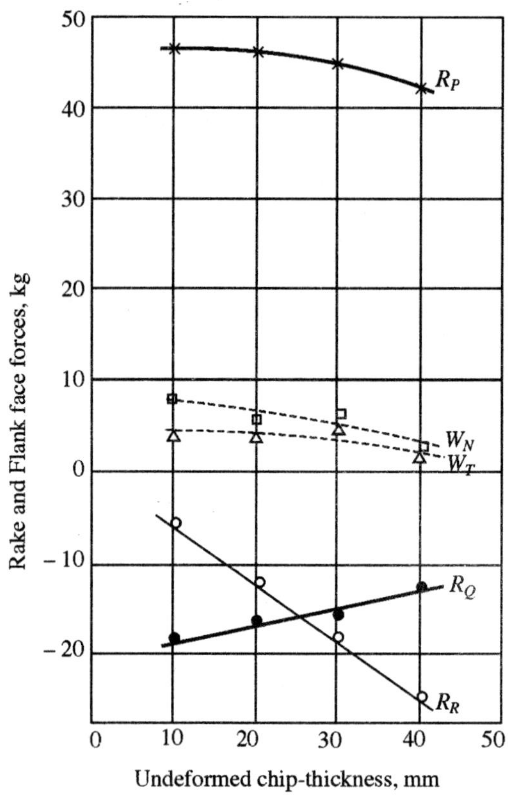

Fig. 5.9 Rake and flank face forces (After Lal, 1982)

The above method cannot be used for evaluating the flank face forces during orthogonal cutting since only four equations for equilibrium of forces can be obtained while six equations are required to obtain the six unknowns. Two more equations could have been obtained in the form of moment equations but the exact line of action of the force vectors are not known.

CHAPTER 6

Measurement of Cutting Forces

6.1 INTRODUCTION

The metal-cutting process can be analysed only after certain observations are made before, during and after a cut. The observations that can be made during cutting are rather limited, one of the most important being the measurement of cutting forces. The knowledge of the magnitude and variation of cutting forces under various cutting conditions is particularly useful for rational design of machine tools. It helps the designer in deciding how the machine should be constructed and what dimensions the various machine components should have, and what power is necessary to drive the tool and the machine table. Its knowledge is also useful in deciding which tool or grinding wheel is suitable for a particular job with regard to its capacity and durability.

The idea of measuring cutting forces occurred to investigators almost at the beginning of 20th century but earlier devices were comparatively crude. Schlesinger (1907) was probably the first to measure these forces from power measurement. Since then many precision force measuring devices (dynamometers) have been developed and successfully used for measuring the cutting forces. Most of these designs, however, make use of bonded wire resistance strain gauges. We shall, therefore, confine our discussion to strain gauge applications. Before discussing the actual cutting force measuring dynamometers, let us consider a few basic methods for measuring forces with the help of strain gauges.

6.2 BASIC METHODS FOR MEASURING FORCES

6.2.1 Axially Loaded Member

Probably the simplest method of measuring unidirectional force is to use some kind of a bar in tension or compression as shown in Fig. 6.1. For maximum sensitivity, strain gauges may be arranged as shown and connected to form a Wheatstone bridge. It is clear that gauges 1 and 4 will measure axial strain while the circumferential strain caused by Poisson's effect will be measured by gauges 2 and 3. The strain in the gauges will be

$$e_1 = e_4 = \frac{F}{AE} \qquad \text{...(6.1)}$$

and

$$e_2 = e_3 = \frac{\nu F}{AE} \qquad \text{...(6.2)}$$

where F is the axial force, A is the cross-sectional area, E is the Young's modulus of elasticity and ν is the Poisson's ratio. Subscripts 1, 2, etc. refer to the gauges. In this arrangement, the spring constant, K_s, is

$$K_s = \frac{AE}{L} \quad ...(6.3)$$

where L is the length of the bar.

With this arrangement of gauges, the bridge output will be insensitive to any loading other than F. We are of course assuming that the gauges have been symmetrically placed. Suppose a bending load causes additional strain in gauge 1, gauge 4 will then have an equal and opposite strain and the effect will cancel out in the bridge since gauges 1 and 4 are in opposite arms. In actual practice, however, they never cancel out completely and the degree of accuracy depends upon the characteristics of the strain gauges and the mounting technique used.

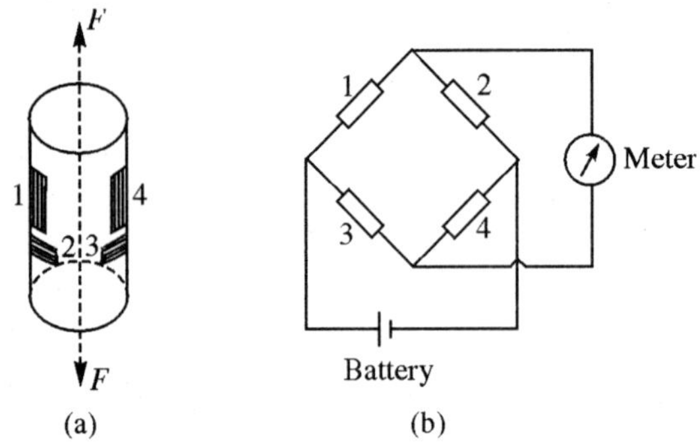

Fig. 6.1 (a) Strain-gauge arrangement for measuring axial load (b) Wheatstone bridge circuit

6.2.2 Cantilever Beam

Cantilever beams are best suited for measurement of bending moment. When the point of application of load is known, the bending moment can be directly converted into force. Consider a rectangular cantilever beam of length L, width b and depth h subjected to forces F_p and F_q as shown in Fig. 6.2(a). It is clear that F_q is measured by gauges 1, 2, 3 and 4 and gauges 1′, 2′, 3′ and 4′ measure F_p. These gauges can now be connected as shown to form two independent Wheatstone bridges for measuring F_q and F_p. If the gauges are mounted symmetrically and properly located with respect to the neutral axis of the beam, the two mutually perpendicular force components F_q and F_p can be measured independently. The strain measured by the gauges are

$$e_1 = e_2 = e_3 = e_4 = \frac{6Fql}{Ebh^2} \quad ...(6.4)$$

and the spring constant will be

$$K_s = \frac{Ebh^2}{4L^3} \quad ...(6.5)$$

Here l is the distance of gauges from the force end. Similarly,

$$e_1' = e_2' = e_3' = e_4' = \frac{6F_p l}{Ehb^2} \quad ...(6.6)$$

and

$$K_s = \frac{Ehb^3}{4L^3} \tag{6.7}$$

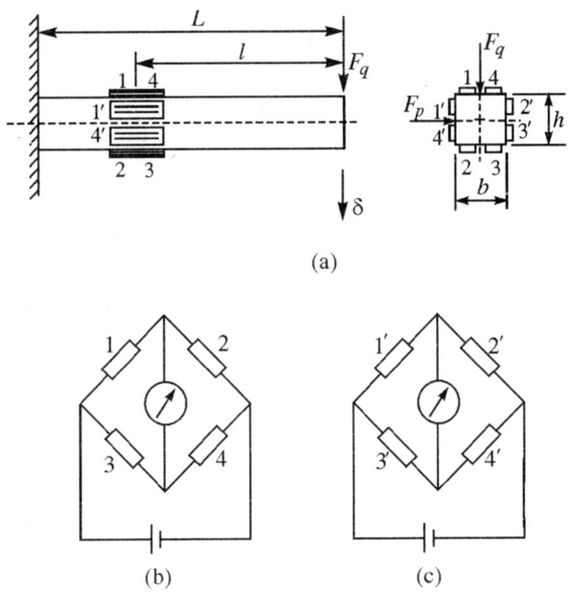

Fig. 6.2(a) Strain-gauge arrangement for measuring F_p and F_q components (b) and (c) are Wheatstone bridge circuits for measuring F_q and F_p, respectively

6.2.3 Rings

Rings have been used for many years for measuring force. It provides a high ratio of sensitivity to stiffness and at the same time has adequate stability against buckling. Consider a thin metallic ring of mean radius r, thickness t, and width b, fixed at the bottom as shown in Fig. 6.3(a). When this ring is subjected to normal force F_q, it will deform as shown in Fig. 6.3(b). The fact that inside of the ring will be subjected to opposite state of strain from the outside allows four active strain gauges to be located at 1, 2, 3 and 4. These form the four arms of the Wheatstone bridge (Fig. 6.3c) and measure force F_q. Thin ring elastic theory gives the strain in these gauges to be

$$e_{90^\circ} = \pm 1.09\left[\frac{F_q r}{Ebt^2}\right] \quad ...(6.8)$$

while the strain in gauges 1′, 2′, 3′ and 4′ located at 39.6° from the vertical axis is zero. When the tangential force F_p only is applied, the ring will deform as shown in Fig. 6.3(d) and strain gauges located at 1′, 2′, 3′ and 4′ form the bridge as shown in Fig. 6.3(e) to measure F_p. The strain in gauges 1, 2, 3 and 4 is zero while the strain in gauges 1′, 2′, 3′ and 4′ is

$$e_{39.6^\circ} = \pm 2.31\left[\frac{F_p r}{Ebt^2}\right] \quad ...(6.9)$$

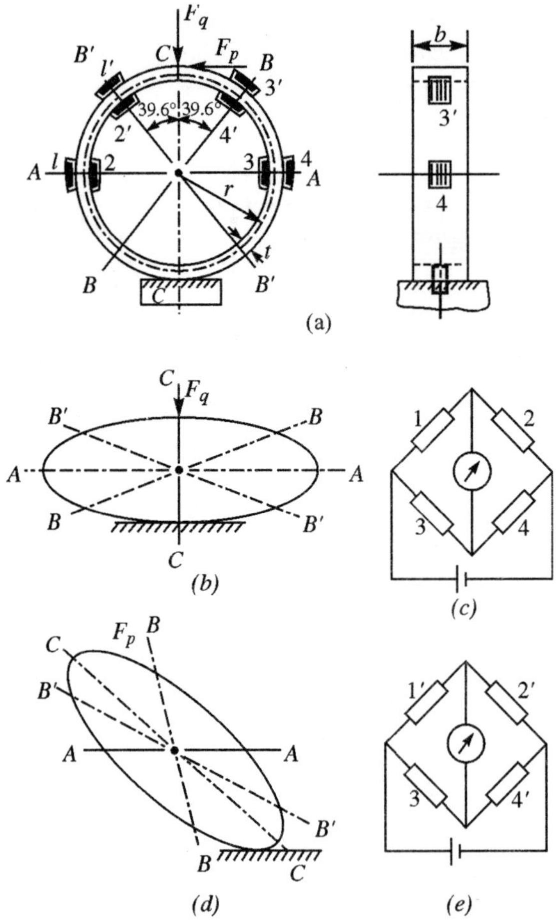

Fig. 6.3 (*a*) Strain-gauge arrangement for measuring F_p and F_q (*b*) Deformation of ring under F_q (*c*) Bridge circuit for measuring F_q (*d*) Deformation of ring under F_p (*e*) Bridge circuit for measuring F_p

Thus, by mounting strain gauges at the inside and outside surfaces of a ring at 39.6° and 90° from the vertical axis *C-C* (Fig. 6.3a), it is possible to measure both normal and tangential components independently. The corresponding stiffnesses are

$$[K_s]_q = \left[\frac{Ebt^3}{1.8r^3}\right] \qquad \text{...(6.10)}$$

and

$$[K_s]_p = \left[\frac{Ebt^3}{9.42r^3}\right] \qquad \text{...(6.11)}$$

There are many practical difficulties which give too little rigidity to these rings in the horizontal direction. These rings always have a tendency to roll. They can, however, be made much more useful by modifying its shape to an octagon as shown in Fig. 6.4. Photoelastic examinations of such rings reveal that the strain nodes now occur at 50° and 90° from the vertical axis. The centrally mounted strain gauges measure F_q as before, while gauges at 50° from the vertical axis pick-up F_p. The strain and stiffness in the two directions are approximately given by the following equations:

$$e_{50^\circ} \approx \pm 1.4 \left[\frac{F_p r}{E b t^2} \right] \quad ...(6.12)$$

$$e_{50^\circ} \approx \pm 0.7 \left[\frac{F_p r}{E b t^2} \right] \quad ...(6.13)$$

$$[K_s]_q \approx \pm \left[\frac{E b t^3}{r^3} \right] \quad ...(6.14)$$

$$[K_s]_p \approx \pm \left[\frac{E b t^3}{3.7 r^3} \right] \quad ...(6.15)$$

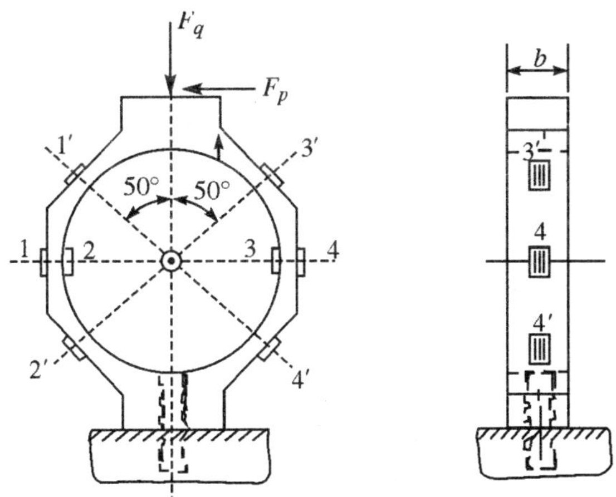

Fig. 6.4 Strain-gauge arrangement on octagonal ring for measuring F_p and F_q

6.3 DYNAMOMETER REQUIREMENTS

There are many practical difficulties in designing a suitable dynamometer. The general requirements for machine tool dynamometers could be summarized as follows:

(*a*) Sensitivity

Determination of cutting force components depends on the measurement of small strains or displacements. Therefore, for high sensitivity these measurements should be as large as possible. But these displacements should not be too large, otherwise this will interfere with the true cutting condition.

(*b*) Rigidity

The dynamometer should be sufficiently rigid, relative to the rest of the machine tool so that it does not affect the cutting condition by causing excessive vibration and chatter. The introduction of the dynamometer would inevitably introduce some flexibility, but this should be restricted to a minimum. In general, the smaller the rigidity, the lower would be the natural frequency and in order that the interference with cutting condition be small, the natural frequency of the dynamometer should be several times the cutting frequency. It is clear that the rigidity requirements conflict with those of sensitivity.

(*c*) Linearity

The relationship between the magnitude of the cutting forces and the deflection should preferably be linear with no hysterisis effects.

(*d*) Interaction

There must be negligible cross-sensitivity among the force components, *i.e.*, one component of the cutting force should not influence the values of the other components. When mutual interference among the force components exist, determination of these components requires the solution of simultaneous equations.

(*e*) Stability

The dynamometer, its electrical circuits etc. must be stable relative to time, temperature and humidity.

What distinguishes the machine tool dynamometers from other force measuring devices is the need for measuring force components simulteneously in two or more directions, and the need for maximum possible stiffness so that the machining process is not affected by the presence of the measuring device. The severity of these requirements has given rise to a large number of widely different designs. The dynamometers that are described in the following pages should be useful in illustrating some of the successful designs.

6.4 MACHINE TOOL DYNAMOMETERS

A simple two-component strain gauge dynamometer is shown in Fig. 6.5. The tool is mounted at the end of a rectangular cantilever beam and forces F_p and F_q acting on the tool can be easily measured with the help of two strain-gauge bridge circuits as discussed earlier. This is one of the simplest arrangements.

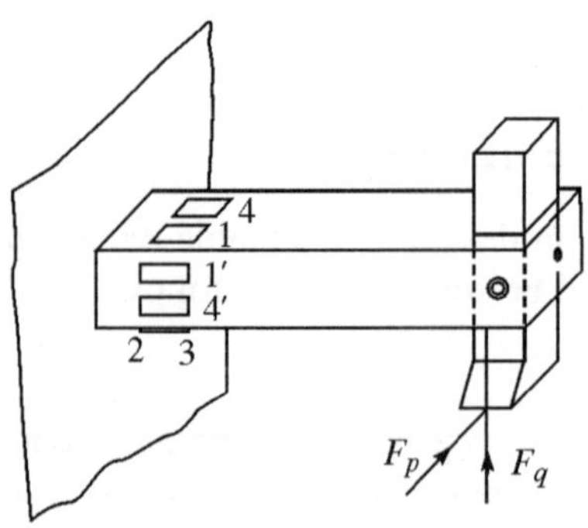

Fig. 6.5 Rectangular cantilever beam tool dynamometer

A strain gauge dynamometer for measuring torque and axial thrust during drilling using bending of arms is shown in Fig. 6.6. Here strain gauges 1 to 8 pickup torque through bending of the arms in the horizontal plane while gauges 9 to 16 form the bridge circuit for measuring axial thrust through bending in the vertical plane.

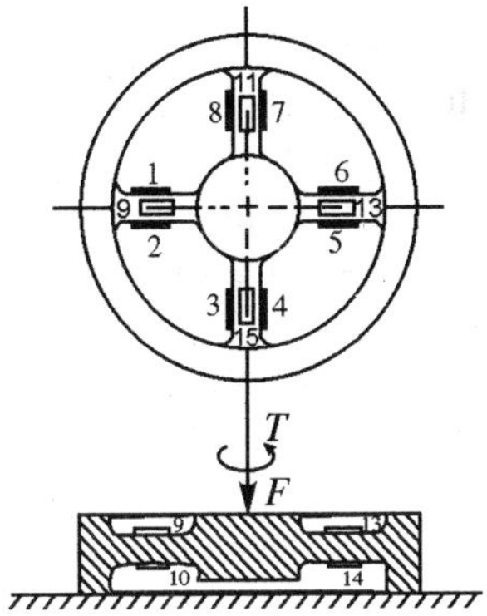

Fig. 6.6 Drilling dynamometer

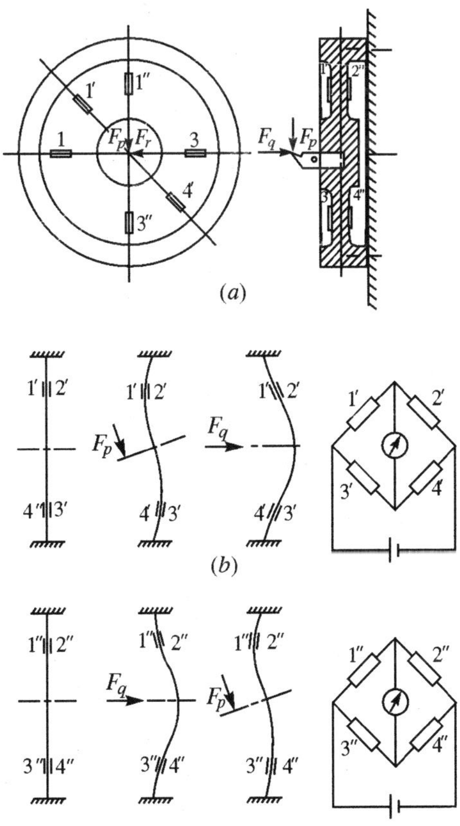

Fig. 6.7 (*a*) Three-component disc dynamometer (*b*) and (*c*) indicate strain gauge arrangements and bridge circuits for measuring F_p and F_q

Figure 6.7(a) shows a tool dynamometer for measuring three force components simultaneously using bending of a disc. The gauge positions for measuring the two horizontal components F_p and F_r as well as the vertical force component F_q are shown. It is clear from Fig. 6.7(b, c) that by suitably connecting the strain gauges, the three force components can be measured with negligible interference.

Circular or octagonal rings, like the ones shown in Figs. 6.3 and 6.4 are not useful by itself but can be made sufficiently stable by extending it as shown in Fig. 6.8. Equations (6.12) to (6.15) can still be used provided the bending of flat top is negligible. Such an extended ring can also be used for measuring moment M, using central strain gauges 1, 2, 3 and 4, but with positions of strain gauges 3 and 4 interchanged. This extended ring unit will give the same results for forces F_p and F_q and their magnitudes will be independent of the position of application as long as they are within the length $2l$. If the force F_q is displayed sideways, the total vertical load carried by the left and the right half rings will be the same, even through the portion of the load carried by either half rings are different. This is because the output from the Wheatstone bridge circuit is proportional to the sum of all strains and hence proportional to the total load. This design is particularly suitable for operations like horizontal milling, horizontal surface grinding, etc. where the workpiece could be mounted on the flat plate of the dynamometer. Alternatively, the cutting tool could be mounted on the dynamometer and two force components can be measured.

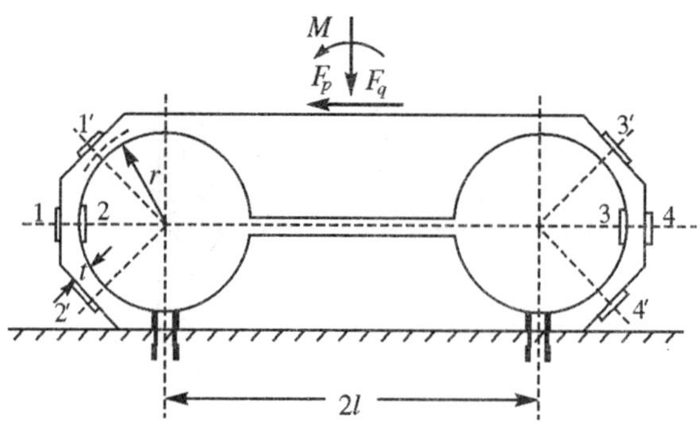

Fig. 6.8 Extended octagonal ring dynamometer

By combining four rings in a form such as shown in Fig. 6.9 and mounting them between two plates, forces in three mutually perpendicular directions as well as the moments around the three axes can be easily measured. If the rings and the gauges are symmetrically located, the forces will be independent of their point of application over a sufficiently large area. This feature is particularly useful when large workpieces are required to be used. In this case, when all four rings are connected to form a single bridge, the vertical force F_q is measured. Rings A and D measure the force component F_r, while rings B and C measure F_p. The moment M can be measured by simply interchanging the arms of F_p or F_r bridges. When moments about the horizontal axes are required to be measured, the gauges in F_q circuit can be rewired. Thus, the device shown in Fig. 6.9 can not only measure the three force components but can also be wired to measure the moments. Dynamometers of this type are useful for studying milling, planing, shaping, drilling and grinding operations.

Another satisfactory three component tool dynamometer using octagonal rings is shown in Fig. 6.10. This measures the force components F_p and F_q as measured by an ordinary extended octagonal ring except that the forces are shared by both the rings. The third component, F_r, puts the elements in the upper ring in tension and those in the lower ring in compression.

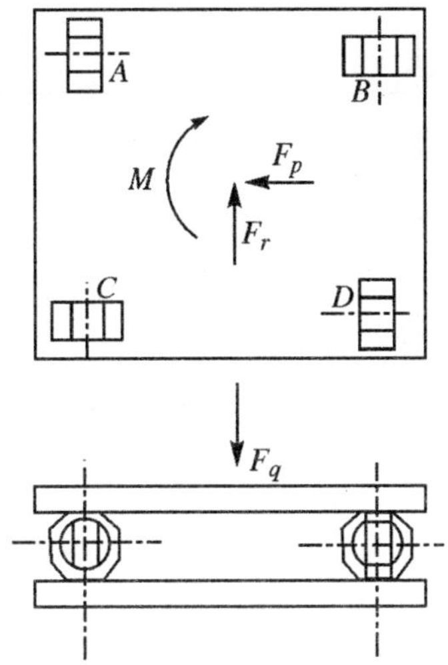

Fig. 6.9 A three-component dynamometer using octagonal rings

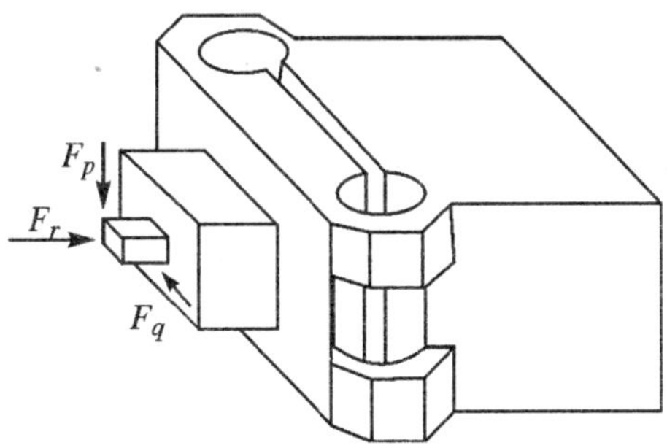

Fig. 6.10 A three-component tool dynamometer

A three-component dynamometer using steel ligaments is shown in Fig. 6.11. It has a workpiece platform supported within a frame by means of four vertical steel ligaments (*a, b, c* and *d*). Horizontal constraints for the platform are provided by two sets of four horizontal ligaments (*e, f, g, h* and *i, j, k, l*) whose ends are attached to the platform and the frame. All attachments are by pin joints, each ligament being pre-stressed by a tensioning screw at one end. The platform is clamped to the centre of the vertical ligaments after these are tensioned. Strain gauges are mounted on both sides of all ligaments and are connected to form three independent bridge circuits thereby allowing independent measurement of the three force components F_p, F_q and F_r.

6.5 DYNAMOMETER CALIBRATION

When calibrating a multi-component dynamometer, there may be cross-interference between the force components, i.e., there may be a fictitious force reading in *y* or *z*-directions when force is applied in the *x*-direction. If this cross-interference is not negligible, cross-sensitivity analysis may be carried out.

Calibration of dynamometers are best accomplished using precision dead weights. Use of lever systems etc. should be avoided since they generally tend to produce errors. When the loads required for calibration are very large, proving rings or a calibrated testing machine should be used.

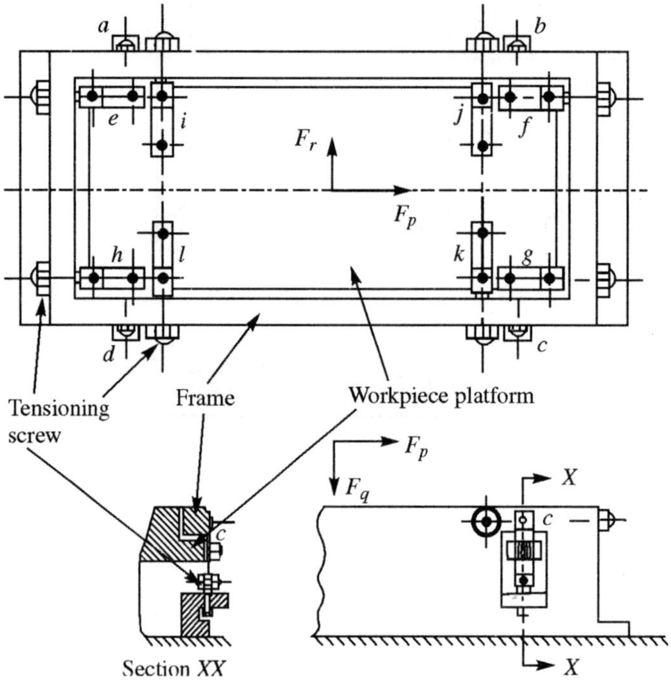

Fig. 6.11 A three-component dynamometer using steel ligaments

The usual method of carrying out static calibration of a dynamometer involves application of load in one direction and recording the outputs (R_x, R_y and R_z) in all directions. Each load cycle is repeated several times in order to evaluate the experimental error. The results of the experimental data obtained in the static calibration will enable the following matrix equation to be written:

$$\begin{bmatrix} R_x \\ R_y \\ R_z \end{bmatrix} = \begin{bmatrix} A_{xx} & A_{xy} & A_{xz} \\ A_{yx} & A_{yy} & A_{yz} \\ A_{zx} & A_{zy} & A_{zz} \end{bmatrix} \cdot \begin{bmatrix} F_x \\ F_y \\ F_z \end{bmatrix} \qquad ...(6.16)$$

By inverting the matrix of the calibration factors it is possible to compute from the values of the outputs of the three channels — the values of three components of the cutting force. Thus

$$\begin{bmatrix} F_x \\ F_y \\ F_z \end{bmatrix} = \begin{bmatrix} A'_{xx} & A'_{xy} & A'_{xz} \\ A'_{yx} & A'_{yy} & A'_{yz} \\ A'_{zx} & A'_{zy} & A'_{zz} \end{bmatrix} \cdot \begin{bmatrix} R_x \\ R_y \\ R_z \end{bmatrix} \qquad ...(6.17)$$

The dynamic calibration is carried out to determine the resonance frequencies of the dynamometer in various force directions. A sinusoidal force (variable frequency) is generally applied along a given direction and output from the dynamometer is recorded on an oscilloscope. The natural frequency in a given direction is obtained when resonance occurs.

6.6 GENERAL REMARKS

The strain gauge translates small changes in dimensions into an equivalent change of resistance. The relationship in terms of strain e is

$$e = \frac{l}{G}\frac{\Delta R}{R} \quad \text{...(6.18)}$$

The values of gauge factor G and resistance R are supplied by the manufacturer and the user determines the change in resistance ΔR corresponding to the input situation. Since the change in the resistance seldom exceeds 0.5% of the gauge resistance, the practical method for measuring this change is by use of a Wheatstone bridge circuit. It is obvious from Fig. 6.12 that for balance

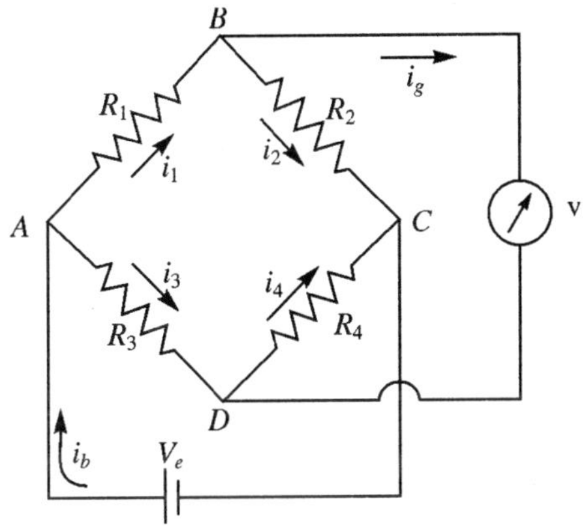

Fig. 6.12 Wheatstone bridge circuit

$$i_g = 0 \quad \text{...(6.19)}$$

and

$$\frac{R_1}{R_2} = \frac{R_3}{R_4} \quad \text{...(6.20)}$$

If we assume that

$$R_1 = R_2 = R_3 = R_4 = R$$

and that the resistance in one of the arms changes by ΔR then the change in the output voltage ΔV_e is given by

$$\Delta V_e = \left[\frac{\Delta R / R}{2(\Delta R / R) + 4}\right] V_e \quad \text{...(6.21)}$$

where V_e is the input voltage. This shows that the behaviour is inherently non-linear. The resistance change in most strain gauge bridge circuits is, however, so small that the behaviour may be assumed to be linear.

It is clear from equation (6.21) that the sensitivity of the bridge circuit can be increased by increasing the impressed voltage V_e. Too high a voltage, however, causes zero drift. Therefore, for stability the input voltage must be low. The maximum permissible voltage can be roughly estimated from

$$V_e = 2\ (WR)^{1/2} \qquad ...(6.22)$$

where W is maximum wattage that a gauge can dissipate. It is clear that use of two or more gauges in each arm in place of one, enables the impressed voltage to be increased without over-heating the gauges.

Assume that a change in resistance by one unit ($+\ \Delta R$) in one of the gauges, say R_1, in Fig. 6.12 increases the output voltage by one unit ($+\ \Delta V_e$). If R_2 also changes by $-\ \Delta R$, then the output voltage will change by $+ 2\ \Delta V_e$. If R_3 and R_4 also change by $-\ \Delta R$ and $+\Delta R$, respectively, then the voltage difference will be $+ 4\ \Delta V_e$. Clearly, the sensitivity can be increased four folds when two gauges are in tension while the other two are in compression. When four active gauges are used, the bridge is also insensitive to temperature changes.

Assuming linear relationship between output voltage and resistance changes, the output voltage v' can be evaluated from

$$v' = \frac{n}{4} G V_e e \qquad (6.23)$$

where n is the number of active arms in the bridge circuit.

The input voltage can be either *AC* or *DC*. *AC* circuit has the advantage of eliminating the possibility of thermoelectric effects and also offers a greater range of stable amplification. AC input must, however, be at a much higher frequency than the force variation to be measured. DC input imposes no such frequency limitations but amplification is more difficult. Further, DC circuits require practicaly no shielding. DC circuits are generally preferred when use of nulling system is required.

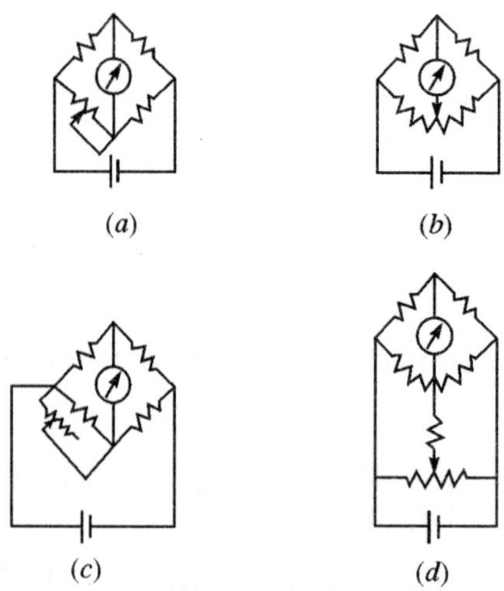

Fig. 6.13 Bridge balance (*a*) Series balance (*b*) Differential series balance (*c*) Shunt balance (*d*) Differential shunt balance

The arrangements for balancing the bridge are shown in Fig. 6.13.

Sometimes capacitors connected across the output terminals as shown in Fig. 6.14 are very effective in eliminating super imposed electrical signals. These capacitors offer a low impedance path and work as an effective short circuit for AC. For estimating the capacity of these capacitors, the vibration signal may be assumed to be an AC signal with frequency equal to the natural frequency of the dynamometer. Introduction of these capacitors in the circuit will not affect the DC output of the bridge.

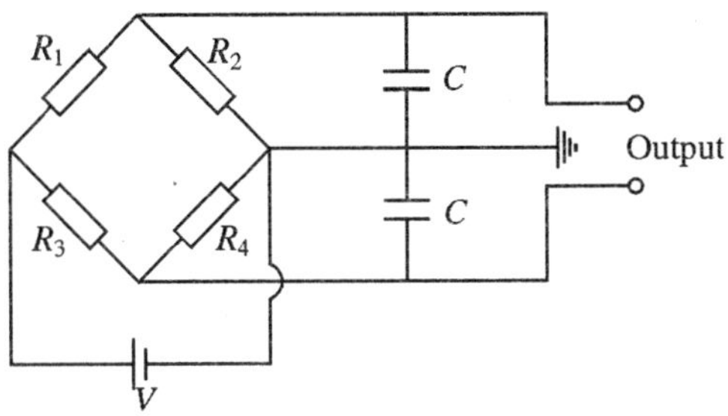

Fig. 6.14 Bridge circuit with capacitors connected across the output terminals

CHAPTER 7

Thermal Aspects of Machining

7.1 INTRODUCTION

During machining of metals, considerable heat is generated when conservation of mechanical energy takes place. This occurs in the following three distinct regions as shown in Fig. 7.1:

(*a*) The shear zone where heat is generated due to plastic deformation of workpiece material.

(*b*) The chip-tool interface zone where heat is generated due to frictional rubbing between the rake face of the tool and chip.

(*c*) The work-tool interface zone where heat is generated due to frictional rubbing between the flank face of the tool and workpiece.

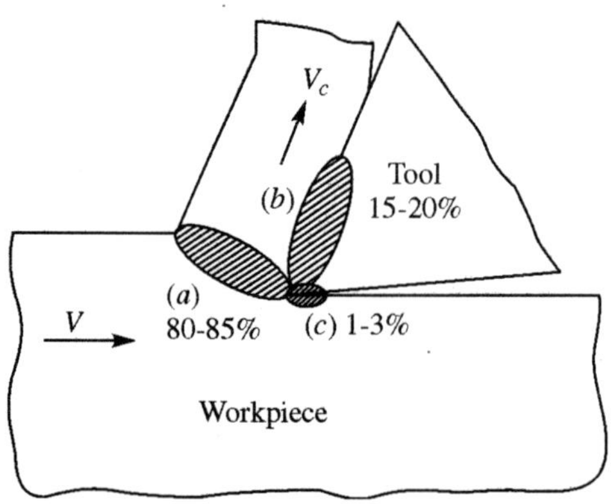

Fig. 7.1 Source of heat in machining

About 80–85% of the heat is generated in the shear zone, while chip-tool and tool-work interfaces contribute 15-20% and 1-3%, respectively. Experiments show that about 80% of the heat generated is carried away by the chip, 15-20% flows into the tool, and less than 5% is conducted into the workpiece. The portion of heat that flows into the tool causes very high temperature in the vicinity of tool tip which in turn decreases the hardness of tool material and in extreme cases may even cause melting. The wear rate of the tool, therefore, increases resulting in a decrease in the useful life of the tool. The thermal stresses developed due to high temperature also contribute to the failure of the tool.

Cutting speed is the most important variable affecting the cutting temperature, while cutting feed is the next important variable, followed by depth of cut. Experiments show that the chip-tool interface temperature increases approximately parabolically with increase in cutting speed (Fig. 2.13). Feed and depth of cut do not have such pronounced effects. The tool temperature is also influenced by the properties of workpiece material since they affect the size of the plastically deformed zone (shear zone) and the chip-tool contact area over which the heat is distributed. The tool geometry, cutting fluids, etc. are other factors that influence cutting temperature.

In order to understand how the heat generated during metal cutting is removed by the chip, tool and workpiece, it is necessary to understand the transfer of heat in a material which is moving relative to the heat source.

7.2 EQUATIONS OF HEAT FLOW

Consider a two-dimensional heat flow problem. The element $ABCD$ through which the heated material flows is shown in Fig. 7.2. The co-ordinates of A is (x,y) and the material at this point is assumed to be at an instantaneous temperature θ. The co-ordinates and temperatures at the other corner points B, C and D are also shown in the figure. The material velocities in x and y-directions are V_x and V_y as shown. Heat is transferred across the boundaries AB, BC, CD, and DA by conduction, because of the temperature gradients, and by transportation because of the flow of heated material across these boundaries.

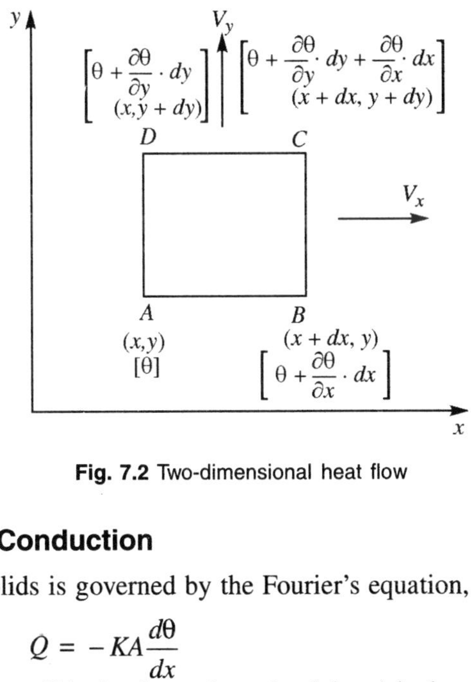

Fig. 7.2 Two-dimensional heat flow

7.2.1 Heat Flow Due to Conduction

The conduction of heat in solids is governed by the Fourier's equation,

$$Q = -KA\frac{d\theta}{dx} \qquad ...(7.1)$$

where Q is the rate of heat flow, K is the thermal conductivity, A is the area of cross-section normal to the x-direction and $d\theta/dx$ is the temperature gradient in the x-direction.

For the element of dimensions (x, y, l) shown in Fig. 7.2, the Fourier's law of heat conduction gives,

$$Q_x = Kdy\frac{\partial\theta}{\partial x} \qquad ...(7.2)$$

and

$$Q_{(x+dx)} = -Kdy\frac{\partial}{\partial x}\left(\theta + \frac{\partial\theta}{\partial x}dx\right) \qquad ..(7.3)$$

where Q_x is the rate of heat flow into the element in x-direction and $Q_{(x+dx)}$ is the rate of heat flowing out. Similarly, in the y-direction,

$$Q_y = -Kdx\frac{\partial\theta}{\partial y} \quad ...(7.4)$$

and

$$Q_{(y+dy)} = -Kdx\frac{\partial}{\partial y}\left(\theta+\frac{\partial\theta}{\partial y}dy\right) \quad ...(7.5)$$

assuming the material to be isotropic with respect to its thermal conductivity, *i.e.*,

$$K_x = K_y = K_z = K \quad ...(7.6)$$

7.2.2 Heat Flow due to Transportation

Heat transferred across any cross-section due to transportation (flow of heated material) Q' can be estimated from,

$$Q' = \rho c V A \bar{\theta} \quad ...(7.7.)$$

where ρ is the density, c is the specific heat, V is the velocity and $\bar{\theta}$ is the average temperature. Thus, heat flow due to transportation along x-direction can be calculated from

$$Q'_x = \rho c V_x\left(\theta+\frac{1}{2}\frac{\partial\theta}{\partial y}dy\right)dy \quad ...(7.8)$$

and

$$Q'_{(x+dx)} = \rho c V_x\left(\theta+\frac{\partial\theta}{\partial x}dx+\frac{1}{2}\frac{\partial\theta}{\partial y}dy\right)dy \quad ...(7.9)$$

Similarly, in the y-direction,

$$Q'_y = \rho c V_y\left(\theta+\frac{1}{2}\frac{\partial\theta}{\partial x}dx\right)dx \quad ...(7.10)$$

and

$$Q'_{(y+dy)} = \rho c V_y\left(\theta+\frac{\partial\theta}{\partial y}dy+\frac{1}{2}\frac{\partial\theta}{\partial x}dx\right)dx \quad ...(7.11)$$

7.2.3 Heat Absorbed

The rate of heat absorbed by the element Q_a can be evaluated from

$$Q_a = \rho(dx\times dy)c\frac{\partial\theta}{\partial\tau} \quad ...(7.12)$$

where τ is time.

7.2.4 Heat Generated

The rate of heat generated in the element

$$Q_r = q_r\,dxdy \quad ...(7.13)$$

where q_r is the rate of heat generated per unit area.

The balance of rate of heat flow, therefore, gives

$$[Q_x - Q_{(x+dx)}]+[Q_y - Q_{(y+dy)}]+ [Q'_x - Q'_{(x+dx)}]+[Q'_y - Q'_{(y+dy)}]+Qr=Qa \quad ..(7.14)$$

Substitution of various values and simplification gives the heat flow equation as

$$\frac{K}{\rho c}\left(\frac{\partial^2\theta}{\partial x^2}+\frac{\partial^2\theta}{\partial y^2}\right)-\left(V_x\frac{\partial\theta}{\partial x}+V_y\frac{\partial\theta}{\partial y}\right)+\frac{q_r}{\rho c}=\frac{\partial\theta}{\partial\tau} \quad ...(7.15)$$

Solution of equation (7.15) is difficult, except for some simple boundary conditions.

7.3 TEMPERATURES IN ORTHOGONAL CUTTING

The region of interest from the point of view of temperature in metal cutting are the shear plane and chip-tool interface temperatures. Consider the case of orthogonal cutting with continuous chips (Fig. 7.3). Here the material approaching the heat source (shear plane *AB)* is heated rapidly to its maximum temperature and then remains at almost constant temperature. The temperature at various points can, therefore, be assumed to reach a steady-state value after cutting for a while and heat flow equation (7.15) can be applied.

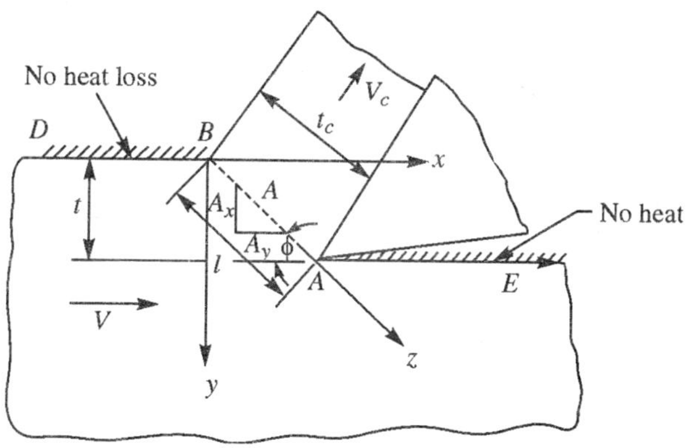

Fig. 7.3 Idealized model for evaluating shear plane temperature distribution

7.3.1 Shear Plane Temperature Distribution

Several attempts have been made to evaluate the shear plane temperature analytically, one of the successful attempts being due to Weiner [1955]. He was able to solve equation (7.15) using the following assumptions for evaluating the shear plane temperature:

(*a*) Heat released at the shear plane is uniform.

(*b*) Workpiece is of infinite mass.

(*c*) Thermal properties of the workpiece and tool materials are independent of temperature.

(*d*) Heat conduction at the tool-work interface is negligible.

(*e*) There is no loss of heat from the workpiece free surfaces *AE* and *BD*, *i.e.*, the free surfaces are insulated.

Taking x and y-directions as shown in Fig. 7.3, the boundary conditions in the workpiece reduce equation (7.15) to

$$\frac{K}{\rho c}\left(\frac{\partial^2\theta_s}{\partial x^2}+\frac{\partial^2\theta_s}{\partial y^2}\right)-V\frac{\partial\theta_s}{\partial x}=0 \quad ...(7.16)$$

where θ_s is the shear plane temperature and V is the cutting speed.

Assuming that no heat is conducted in the material in the direction of its motion, the transfer of heat being mainly by transportation, the conduction term in the direction of motion can be neglected and equation (7.16) becomes,

$$\frac{K}{\rho c}\frac{\partial^2\theta_s}{\partial y^2}-V\frac{\partial\theta_s}{\partial x}=0 \quad ...(7.17)$$

Weiner was able to solve this equation within the stipulated boundary conditions and gave the following equation for the shear plane temperature distribution:

$$\theta_s = \frac{q_s}{\rho c V\sin\phi}\left[(1-2B^2x).erf(B\sqrt{x})+\frac{2B}{\sqrt{\pi}}.\sqrt{x}.\exp(-B^2x)-2B^2x\right] \quad ...(7.18)$$

where

$$B=\frac{1}{2}\tan\phi.\sqrt{\frac{V}{t}} \quad ...(7.19)$$

q_s is the heat released at the shear plane, t is the undeformed chip thickness and ϕ is the shear angle.

7.3.2 Temperature Distribution in Chip

Several methods are available for evaluating the chip-tool interface temperature analytically. Rapier [1954] was able to obtain a solution for the chip-tool interface temperature using equation (7.15). He assumed that the heat source due to friction between the chip and the tool was a plane heat source of uniform strength. The boundary conditions were taken as follows:

(*a*) The material leaves the shear plane *AB* (Fig. 7.4) at a uniform temperature θ_s.

(*b*) The frictional heat source along the chip-tool interface *AC* is uniform.

(*c*) There is no heat loss from the chip surfaces *CG* and *BH*.

(*d*) As $x \to \infty$, the temperature across the chip θ_c becomes uniform, *i.e.*,

$$\theta_c = \theta_s + \theta_f \quad ...(7.20)$$

where θ_f is the temperature increment in the chip due to friction at the chip-tool interface.

As before, it is assumed that heat conduction at the tool-work interface is negligible. Further, the thermal properties of work and tool materials are assumed to be independent of working temperature.

Taking x and y directions as shown in Fig. 7.4 and using the above boundary conditions, equation (7.15) reduces to

$$\frac{\partial^2\theta_c}{\partial y^2} = \frac{\rho c}{K}\frac{\partial\theta_c}{\partial\tau} \quad ...(7.21)$$

since transfer of heat along the x-direction may be assumed to be mainly by transportation and no heat is generated ($q_r = 0$).

Rapier (1954) was able to solve this equation within the stipulated boundary conditions and gave the following equation for evaluating the temperature distribution in the chip:

$$\frac{a}{R_T}\left(\frac{\theta_c - \theta_s}{\theta_f}\right) = \frac{(t_c - y)^2}{2t_c^2} + \frac{x}{R_T\, t_c} - \frac{1}{6} + \frac{2}{\pi^2}\sum_{n=1}^{n=\infty}\left[\exp\left(-\frac{n^2\pi^2 x}{R_T t_c}\right)\cos\left(\frac{\pi n y}{t_c}\right)/n^2\right] \quad ...(7.22)$$

Here R_T is the thermal number which can be evaluated from,

$$R_T = \frac{\rho c V_c\, a t_c}{K} \quad ...(7.23)$$

Where V_c and t_c are the chip velocity and chip thickness and the chip-tool contact length AC = (constant). $t_c = at_c$.

For temperature distribution at the chip-tool interface, *i.e.*, for values of $y = 0$, Rapier [1954] obtained the following approximate solution:

$$\frac{a}{R_T}\left[\frac{\theta_c - \theta_s}{\theta_f}\right] \approx 1.13\sqrt{\frac{x}{R_T t_c}} \quad ...(7.24)$$

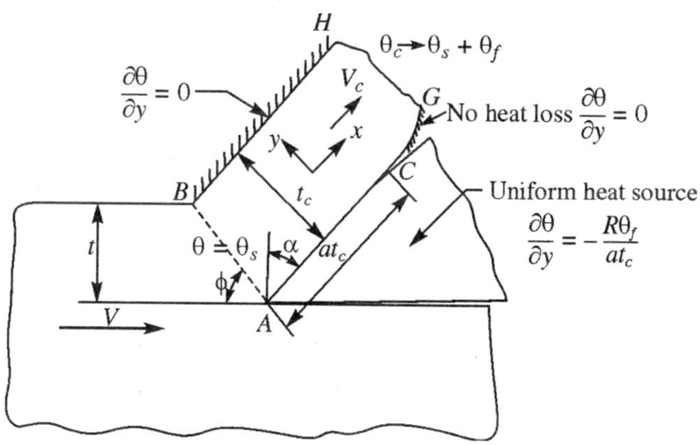

Fig. 7.4 Idealized model for evaluating the chip-tool interface temperature

7.3.3 Average Shear Plane Temperature

The average temperature at the shear plane $\bar{\theta}_s$ can be estimated from the rate at which shear energy (W_s) is expended along the shear plane. Thus,

$$W_s = F_s\, V_s \quad ...(7.25)$$

where F_s is the shear force and V_s is the shear velocity. During plastic deformation, almost all the energy used is converted into heat. Since metal cutting is a process of plastic deformation, it may be assumed that all the energy is converted into heat. Further, most of the heat generated (about 80%) is carried away by the chip. Thus, for orthogonal cutting,

$$\rho t b c V(\bar{\theta}_s - \theta_o) = \beta W_s = \beta F_s V_s \quad ...(7.26)$$

where θ_o is the ambient temperature, β is fraction of heat going into the chip ($\approx$ 0.8), and b is the width of the workpiece. Therefore,

$$\bar{\theta}_s = \frac{\beta F_s V_s}{\rho t b c V} + \theta_o \quad ...(7.27)$$

For orthogonal cutting conditions F_s and V_s can be estimated from equations (4.14) and 4.21) as

$$F_s = F_p \cos\phi - F_q \sin\phi$$

and

$$V_s = \frac{\cos\alpha}{\cos(\phi - \alpha)} V$$

where F_p and F_q are the cutting and thrust forces, and α is the rake angle. Substituting in equation (7.27), the average shear plane temperature for orthogonal cutting can be obtained as

$$\bar{\theta}_s = \frac{\beta(F_p \cos\phi - F_q \sin\phi)\cos\alpha}{\rho t b c \cos(\phi - \alpha)} + \theta_o \quad ...(7.28)$$

7.3.4 Average Chip-tool Interface Temperature

The plastically deformed metal comes out of the shear zone in the form of a chip. The chip is already heated to an average temperature $\bar{\theta}_s$ and is further heated when it slides along the rake face. The frictional energy W_f at the chip-tool interface can be estimated from,

$$W_f = F\,V_c \quad ...(7.29)$$

where F is the friction force and V_c is the chip velocity. The average temperature rise in the chip due to rubbing can be estimated using Jaeger's solution for mean temperature at the interface of a perfectly insulated slider moving across a conducting surface with a constant velocity. If heat is applied continuously and uniformily to an interface of area $2l' \times 2m'$ as shown in Fig. 7.5, the mean temperature $\bar{\theta}$ can be estimated from,

$$\bar{\theta} = \frac{0.754\, q l'}{K\sqrt{L'}} + \theta_o \quad ...(7.30)$$

where q is the rate of heat dissipation at the interface, K is the thermal conductivity of the stationary surface and

$$L' = \frac{V' l'}{2\alpha_d} \quad ...(7.31)$$

Here V' is the slider velocity in x-direction and α_d is the thermal diffusivity of the conducting material.

Consider the case of chip sliding along the rake face (Fig. 7.4). The frictional energy per unit area q_f can be evaluated as

$$q_f = \frac{F V_c}{(a t_c) b} \quad ...(7.32)$$

Here the width of chip is equal to b, the width of workpiece. Approximately 90% of the frictional energy is carried away by the chip and only, about 10% enters the tool. Thus, q in equation (7.30) may be taken as 0.9 q_f. The parameter L' in this equation is given by

$$L' = \frac{V_c\left(\frac{1}{2}at_c\right)}{2\alpha_d} = \frac{at_c V_c}{4\alpha_d} \qquad ...(7.33)$$

since $l' = \frac{1}{2}at_c$, and $V' = V_c$. The tool is now the stationary surface and the chip is the conducting material.

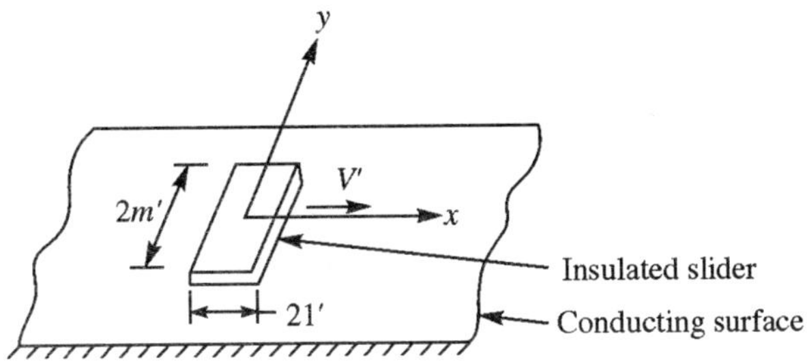

Fig. 7.5 Scheme of moving heat source

The average temperature at the chip-tool interface $\overline{\theta}_f$ can now be estimated as

$$\overline{\theta}_f = \frac{0.754 \times 0.9 q_f \frac{1}{2}at_c}{K\sqrt{\frac{at_c V_c}{4\alpha_d}}} + \overline{\theta}_s$$

or

$$\overline{\theta}_f = \frac{0.678\ q_f \sqrt{at_c}\sqrt{\alpha_d}}{K\sqrt{V_c}} + \overline{\theta}_s$$

or

$$\overline{\theta}_f = \frac{0.678\ F\sqrt{V_c}\sqrt{\alpha_d}}{\sqrt{at_c}\,Kb} + \overline{\theta}_s \qquad ...(7.34)$$

since the chip is already heated to an average temperature $\overline{\theta}_s$. The average shear plane temperature $\overline{\theta}_s$ is known from equation (7.28). Substituting for F and V_c from equations (4.12) and (4.21) as

$$F = F_p \sin\alpha + F_q \cos\alpha$$

and

$$V_c = \frac{\sin\phi}{\cos(\phi - \alpha)}V$$

$\overline{\theta}_f$ can be obtained as

$$\overline{\theta}_f = \frac{0.678.\sqrt{\alpha_d}\ V\sin\phi\,(F_p \sin\alpha + F_q \cos\alpha)}{K\sqrt{at_c}\cos(\phi - \alpha)} + \overline{\theta}_s \qquad ...(7.35)$$

7.4 EXPERIMENTAL DETERMINATION OF CUTTING TEMPERATURES

Several methods have been used for measuring temperatures during metal cutting. These include calorimetric method, thermocouple techniques, use of radiation pyrometers, temperature sensitive chemicals, etc. The technique most widely used for the study of the metal cutting temperatures has been the tool-work thermocouple technique.

7.4.1 Tool-work Thermocouple Technique

The tool-work thermocouple technique is shown schematically in Fig. 7.6. Here the thermocouple is formed at the junction of cutting tool and workpiece with chip-tool interface as the hot junction. The *emf* generated is taken as a measure of temperature in this region. While using this technique it is important to insulate the thermocouple circuit from the rest of the machine. Because of the asperities on the contacting surfaces a large number of point junctions are formed. Each of these individual asperities provide thermoelectric responses acting in parallel and the measured *emf* (E_m) is an average of all these junctions. If the resistance of each of these thermocouple junctions are assumed to be equal, then

$$E_m = \frac{\sum_1^n R_n E_n}{\sum_1^n R_n} = \frac{1}{n}\sum E_n$$

where R_n and E_n are the resistance and *emf* of each of the n junctions.

This technique has been extensively used for investigating the effect of cutting conditions on cutting temperatures. The method, however, gives no indication of the temperature distribution.

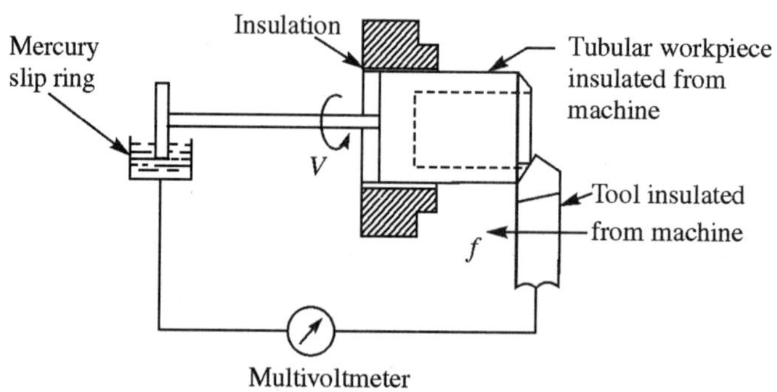

Fig. 7.6 Tool-work thermocouple technique for measurement of cutting temperature

7.4.2 Infrared Photographic Technique

This technique was employed by Boothroyd [1963] for determining the temperature distribution in the cutting zone. The technique involved taking an infrared photograph of the side surface of the cutting zone along with a heated strip of known temperature distribution. The temperature distribution in the heated strip was obtained with the help of thermocouples. By comparing the optical density of the strip photograph with the known temperatures, optical density-temperature calibration was obtained. This calibration was then used to obtain the temperature distribution in the cutting zone.

merely be unsatisfactory for the required purpose. At this point, the tool is resharpened and used again or discarded altogether.

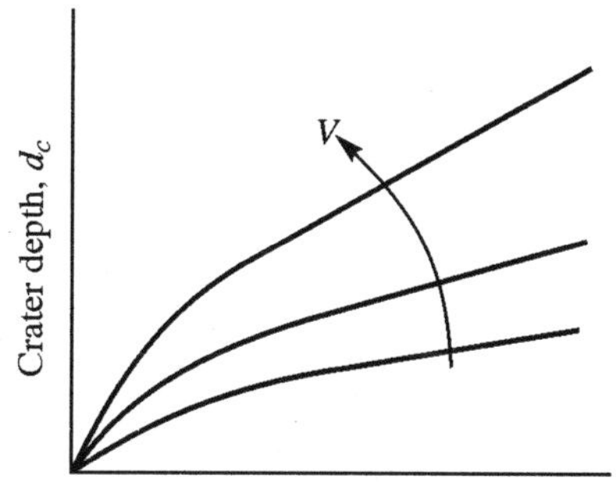

Fig. 8.3 Variation of maximum crater depth with cutting time

There are various ways in which the tool life can be specified, *e.g.*, cutting time to failure, cutting length to failure, volume of material removed to failure, number of components produced to failure, etc. Similarly, there are various ways in which tool failure can be identified, *e.g.*, wear land size, depth of crater, tool wear volume, magnitude of cutting forces, surface finish value, change in component size, total destruction of tool, etc. It is obvious that the tool life value will depend on the tool failure criterion used.

One never waits for the total destruction of tool which may be either due to plastic deformation supported by high temperatures and high stresses, or mechanical fracture caused by excessive forces or impact. Such failures can be prevented by proper selection of tool material and tool geometry. All other tool failure criteria are related to blunting of cutting edge due to unavoidable wearing process. It is, therefore, not surprising that most studies on tool life have been confined to wear aspects.

The virtue of the wear-land size as the criterion for tool failure is that it can be very easily measured. Further, the size of the wear land is related to most of the other parameters which can be used as a failure criterion.

Crater wear has also been used as a tool failure criterion. It is evident that a large crater will weaken the tool and ultimately result in total failure. Several crater parameters have been suggested as the tool failure criterion but the one most commonly used is the maximum depth of crater. From the point of view of measurements, crater depths are much more cumbersome to measure than the wear land. Besides, crater wear is a more realistic tool failure criterion when high-cutting speeds are used.

8.5 VARIABLES AFFECTING TOOL LIFE

The variables that have significant effect on tool life are:

(*a*) the cutting conditions,
(*b*) the tool geometry,
(*c*) the tool material,
(*d*) the work material and,
(*e*) the cutting fluid.

8.5.1 Cutting Conditions

Practical experience has shown that the wear-land size is usually the limiting factor determining the useful life of a cutting tool. When the wear land-cutting time curves are obtained at various cutting speeds (Fig. 8.4), it is found that the rate of wear increases as cutting speed is increased. From curves such as these, the tool life at various cutting speeds for a critical wear-land size l_w* can be obtained. When these values are plotted on logarithmic coordinates (Fig. 8.5), a straight line relationship is obtained and the relationship could be represented by an equation of the form

$$VT^n = C \qquad ...(8.1)$$

where V is the cutting speed, T is the tool life and n and C are constants. Taylor (1907) was the first to show from his experimental work that tool life varied with cutting speed according to this equation and this relationship has come to be known as the Taylor tool life equation.

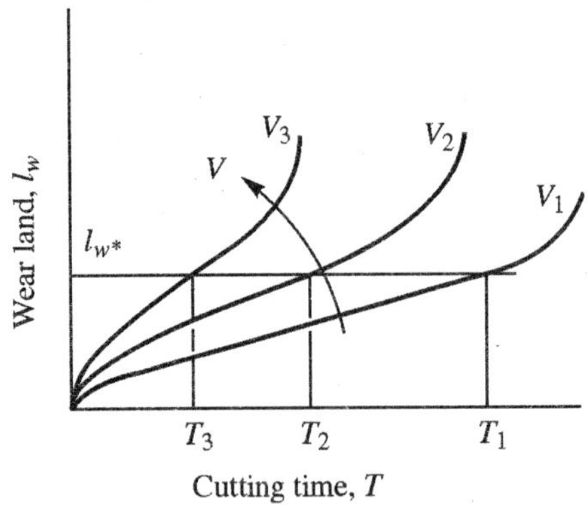

Fig. 8.4 Wear-land growth at different cutting speeds

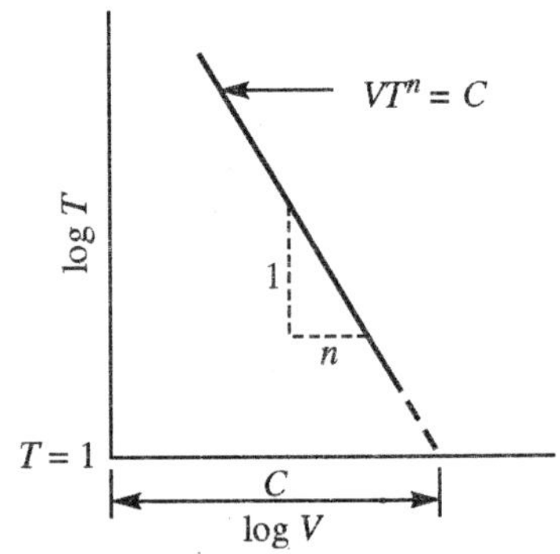

Fig. 8.5 Tool-life curve

The variables speed, feed and depth of cut, affect the material removal rate and hence have a direct bearing on the production cost. Equation (8.1) can be rewritten as

$$T = \frac{C^{1/n}}{V^{1/n}} \qquad ...(8.2)$$

to show the effect of cutting speed. Similar trends occur for feed and depth of cut and tool life T can be expressed as

$$T = \frac{C_1}{V^{1/n} f^{1/n1} d^{1/n2}} \qquad ...(8.3)$$

where f is the feed, d is the depth of cut, C_1 is a constant for a given tool-work material combination and tool geometry, and $1/n_1$ and $1/n_2$ are the exponents for feed and depth of cut, respectively. Equation (8.3) is an extension of the Taylor equation to include the effect of feed and depth of cut. Here $1/n > 1/n_1 > 1/n_2$ and indicates that the cutting speed has the greatest influence on tool life followed by feed and then depth of cut.

The effects of speed, feed and depth of cut on tool life have been explained in terms of cutting temperature. It has been shown experimentally that tool life is related to cutting temperature irrespective of the cutting condition (Fig. 8.6) by a relationship of the form

$$T\theta^m = C_2 \qquad (8.4)$$

where m and C_2 are constants. In general, the cutting temperature θ will rise with increase in V, f and d and hence the effect of these variables can be combined in terms of temperature.

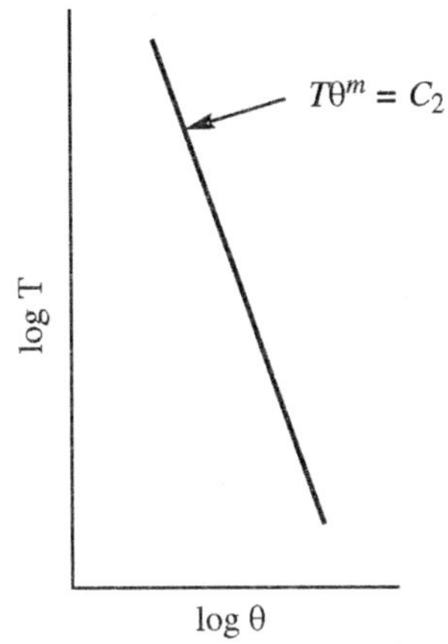

Fig. 8.6 Variation of tool life with cutting temperature

8.5.2 Tool Geometry

A typical relationship showing the variation in tool life with rake angle during orthogonal cutting is shown in Fig. 8.7. Increasing the rake angle reduces the cutting forces and the cutting temperature resulting in increased tool life. When the tool rake angle is large, although the cutting forces are lower, the tool edge is weakened resulting in increased wear due to chipping of the cutting edge. These conditions give an optimum rake angle which gives the maximum tool life. The optimum value of rake angle is, however, a function of the properties of work material. Higher the strength of workpiece material, lower is the value of optimum rake angle.

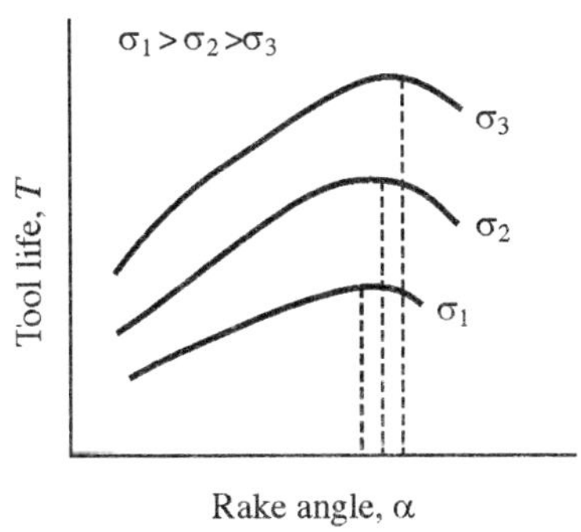

Fig. 8.7 Effect of rake angle on tool life for work materials of different yield strengths

The effect of flank angle on tool life during orthogonal cutting is shown in Fig. 8.8. Here also the tool life increases initially with increase in flank angle and then decreases giving an optimum value. Increasing the flank angle reduces rubbing between the tool and the workpiece and hence improves the tool life. Too high a value, however, weakens the tool and reduces its life. The optimum value of flank angle is also affected by the feed rate–higher the feed rate lower is the optimum value. The flank angle, therefore, should be low if higher feed values are to be used. This is necessary for providing increased strength and better heat dissipation when the feed is increased.

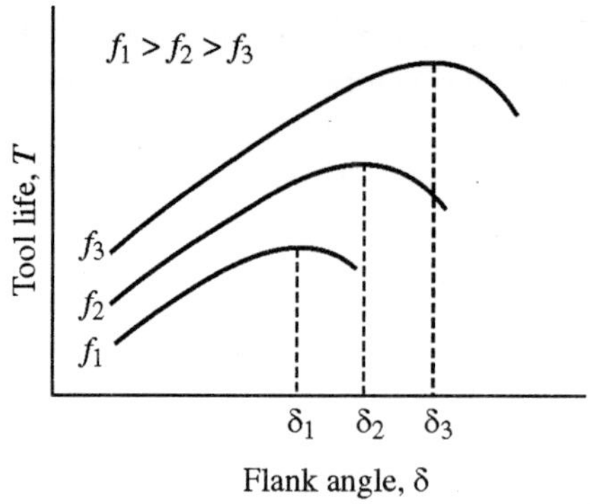

Fig. 8.8 Effect of flank angle on tool life at different feed rates

Similar effect is obtained during turning when the side cutting-edge angle is varied. For a given feed and depth of cut, although the area of cut is not affected by variation in the side cutting-edge angle, the length of the engaged cutting edge increases with increase in the side cutting-edge angle. This gives lower cutting temperature and increases the tool life. Figure 8.9 shows how the optimum value of side cutting edge angle is affected when the strength of workpiece material is varied.

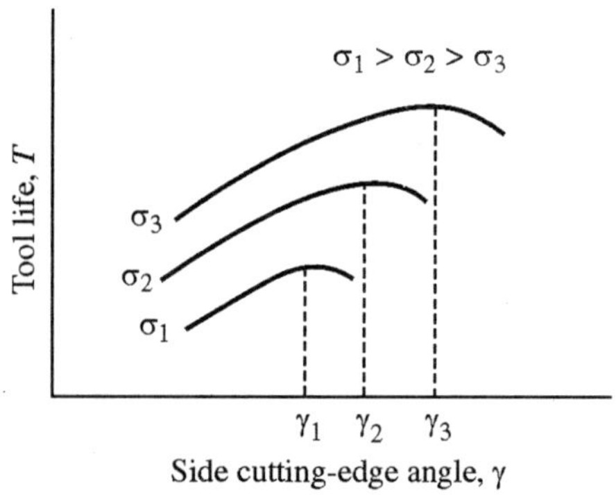

Fig. 8.9 Effect fo side cutting-edge angle on tool life for work materials of different yield strengths

A nose radius on the tool is favourable for long tool life. A nominally sharp tool is highly stressed at the tip and has short tool life. It will also have a tendency to make grooves on the work surface resulting in poor surface finish. An increase in nose radius, therefore, improves the tool life. Increasing nose radius also increases the cutting forces and is conducive to chatter. From these considerations, the nose radii in the range of 0.5-2.5 mm are used most often.

8.5.3 Tool Materials

The basic requirements of cutting tool materials are hardness, toughness, wear and shock resistance and its ability to maintain these properties at cutting temperatures. Hot hardness is necessary to enable the cutting tool to retain its cutting ability, while toughness enables the tool to withstand the forces, absorb shocks and prevent chipping of the cutting edge. These requirements are obviously conflicting and no single tool material will meet all the requiremetns. Other properties that are often considered are hardenability, weldability, grindability, thermal conductivity, coefficient of thermal expansion, etc.

The demand for higher productivity has led to the development of a variety of cutting tool materials with vastly improved properties. Each stage of development has facilitated the use of higher cutting speeds. History, however, reveals that no tool material has been able to fully replace the older one since each one of them has a unique combination of properties. A wide variety of cutting tool materials are still in use and the common ones are:

(*a*) carbon tool steel,

(*b*) high-speed steel,

(*c*) cast non-ferrous alloys,

(*d*) cemented carbides,

(*e*) sintered oxides,

(*f*) artificial abrasives and

(*g*) diamond.

Carbon tool steel containing 0.9 to 1.2% carbon and some alloying elements is perhaps the oldest cutting tool material but has little use in today's manufacturing operations.

High-speed steel (HSS) are alloy steels mainly consisting of about 18% tungsten, 4% chromium and 1% vanadium. In 'Moly'HSS, tungsten is replaced partially by molybdenum and the compositon is generally 8% molybdenum, 4% chromium, 1.5% tungsten and 1% vanadium. In super HSS about 12% cobalt is added for better hot hardness and resistance to abrasion with some sacrifice in toughness.

Cast non-ferrous alloys contain no iron and are cast into their final shape. A typical cast nonferrous alloy contains 43-48% cobalt, 17-19% tungsten, 30-35% chromium and about 2% carbon.

Cemented carbides or sintered carbides are composed of hard metallic carbides bonded together in a metallic matrix. The main ingredients in most common cemented carbide tools are tungsten carbide and cobalt binder. Sometimes titanium or tantalum carbides are added to reduce cratering wear.

Sintered oxides are generally made from sintered aluminium oxides. Sintered alumina containing about 90% aluminium oxide gives best machining results.

Artificial abrasives and diamond are used for grinding and polishing purposes only and their use in metal machining is limited.

UCON and cubic boron nitride (CBN) are two relatively new tool materials that have been introduced in early 70s. UCON consists of 50% columbium (niobium), 30% titanium and 20% tungsten and permits 60% increase in cutting speed when compared with tungsten carbide. CBN has hardness next to diamond and can be used at speeds over five times that of carbides. CBN has also been successfully used for grinding purposes.

Another important development is the introduction of coated carbides for countering crater formation. A thin film ($\approx$ 0.005 mm) of TiC when deposited on cemented carbide tools reduces crater formation due to diffusion of coating material into the tool surface as wear proceeds. Other coatings that have been successfully used are Al_2O_3 and TiN. While TiC and Al_2O_3 provide chemically stable layer between chip and tool, TiN provides low tool friction. Multi-layer coatings of TiC and TiN have also been used. In this arrangement, the stable diffusion barrier is provided by TiC, which prevents iron from entering the cobalt phase, and carbon from leaving the carbide substrate, while TiN provides low chip-tool interface friction.

The most desirable property of a tool material for machining purposes appears to be high hardness that is maintained at cutting temperatures. An increase in hardness, however, decreases the toughness or impact strength. Further, harder the material, more difficult is the problem of giving it the required shape and geometry. In view of these considerations, HSS still appears to be the most popular tool material. Taking hardness as a guide to wear resistance and considering the variation in the hardness of tool material with temperature, the effect on tool life can be explained. Figure 8.10 shows the effect of temperature on the hardness of typical tool materials. High-speed steels, for example, are good upto about 600 °C, beyond which the hardness drops rapidly. Carbon tool steels, on the other hand, can withstand very low temperature, hence are good for cutting at low speeds only. Sintered carbides and cemented oxides retain their hardness at elevated temperature also and hence are suitable for use at high cutting speeds. It has been found experimentally that the value of exponent n is higher for harder tool material. Thus, the effect of cutting speed on tool wear and tool life will be greater when softer tool materials are used.

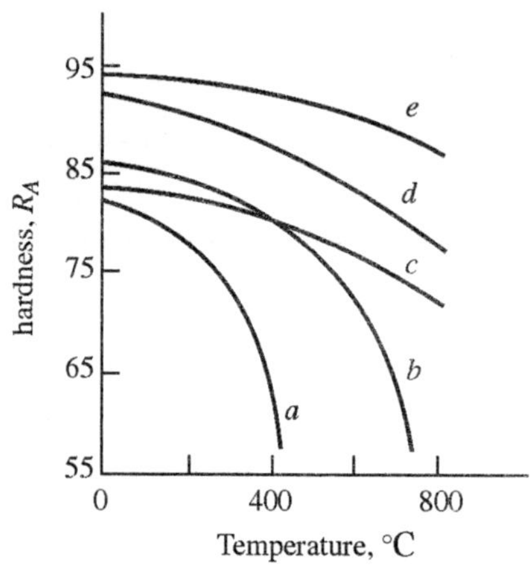

Fig. 8.10 Variation of tool hardness with temperature (a) Carbon tool steels (b) High speed steels (c) Cast non-ferrous alloys (d) Cemented carbides (e) Sintered oxides

8.5.4 Work Materials

The work material, like tool material, also has a significant effect on tool life. In general, harder the work material, higher will be the tool wear rate and lower will be the tool life. It has been shown experimentally that for a constant tool life, the work material hardness and cutting speed are approximately related by an equation of the type

$$V_T = \frac{C'}{H^{n'}} \tag{8.5}$$

where V_T is the cutting speed for a given tool life, C' and n' are constants and H is the hardness of work material.

8.5.5 Cutting Fluids

Most cutting operations are carried out in the presence of a cutting fluid. These fluids are primarily used for decreasing the cutting temperature and the chip-tool interface friction. They also serve to keep the workpiece cool to avoid thermal expansion, promote easy handling, provide a rust-proof layer on the finished surface and remove chips from the machining area.

Cutting fluids are either oil based or waterbased. Oil-based fluids are generally effective in reducing friction in low-speed cutting operations such as tapping, broaching, etc. or in operations such as milling where the tool is repeatedly exposed to the fluid. Thus, their use is generally limited to operations where cutting temperatures are relatively low.

As discussed earlier, the stresses on the rake face of the tool are quite high and in most situations sticking friction prevails near the tool tip. Therefore, the possibility of formation of full fluid film in the chip-tool interface zone is remote. The mode of lubrication is one of boundary lubrication and the chemical property of the cutting fluid is more important than its physical properties. The additives in the cutting fluid generally react with the metal surface to form a low shear strength solid lubricant and prevent welding at the chip-tool interface. This, in turn, reduces the interface friction.

At high cutting speeds, oil-based cutting fluids are ineffective and it is the cooling action that is operative. The water-based fluids are, therefore, more effective because of their high specific heat and thermal conductivity. Water-based fluids, like the oil-based fluids, contain ingredients such as sulphur, chlorine or organic compounds but these are in soluble forms.

By flowing the cutting fluid over the tool, chip and workpiece, the heat generated at the chip-tool and tool-work interfaces as well as in the shear zone is removed thereby decreasing the cutting temperature. This reduction in cutting temperature results in a decrease in wear rate and an increase in tool life. Cutting fluids are generally effective in removing heat from exposed chip, tool and workpiece surfaces and the effect of these heat losses have only a small influence on the cutting zone temperatures. Tool wear, however, is very sensitive to temperature and even a small change in temperature in the cutting region has a pronounced effect on tool wear and tool life.

8.6 DETERMINATION OF TOOL-LIFE EQUATIONS

The determination of tool-life equations present many practical difficulties since the variables involved are numerous and the tests are often time consuming and expensive. In practical situations, however, the number of variables can be reduced from other considerations. The work material is generally known and the tool material, tool geometry and cutting fluid can be selected from handbooks and catalogues. Therefore, the effects of cutting variables-speed, feed and depth of cut, only need to be evaluated. Various methods have been developed for evaluating tool life but here only the conventional tool life test will be described. As mentioned earlier, the cutting speed has the greatest influence on tool life followed by feed and depth of cut.

In order to evaluate tool-life constants n and C of equation (8.1), cutting tests are carried out on a lathe at a suitable feed rate and depth of cut. The workpiece is generally a circular bar but some times tubular specimens are used for ease in measuring the wear-land size (Fig. 8.11). For a particular cutting speed, cutting is interrupted at suitable intervals of time and the length of wear land is measured. The experiment is repeated at several speeds and plots such as those shown in Fig. 8.4 are obtained. Plotting of tool wear-time curve is useful since it helps in deciding about the time intervals to be taken for measuremetns. Once the wear-land growth pattern is established, longer time intervals may be taken to cut down the time to complete the experiments. Tool-life values for a critical wear-land size can now be obtained and plotted on logarithmic co-ordinates to obtain the values of n and C graphically (Fig. 8.5). Alternatively, the exponent n can be evaluated from the relationship

$$n = \frac{\log(V_b / V_a)}{\log(T_a / T_b)} \qquad ...(8.6)$$

where V_a and V_b are any two cutting speeds and T_a and T_b are the corresponding tool life values. The constant C can now be evaluated by substituting the value of n in equation (8.1) and using any set of V and T.

Even in very carefully conducted tests, considerable scatter in the data is obtained and it is necessary to carry out tests at several speeds. This scatter is not only due to the uncertainty in the measurements but also due to the variation in the physical and metallurgical properties of the work material. Experiments indicate that typical values of n for common tool materials are :

HSS	:	0.1 to 0.15
Cast alloys	:	0.15 to 0.2
Cemented carbides	:	0.2 to 0.5
Sintered oxides	:	0.5 to 0.8

When the effects of other variables such as feed and depth of cut are also required, curves such as those shown in Fig. 8.4 may be obtained by repeating the tests at different feeds and depths of cut. Again tool-life values are plotted on logarithmic co-ordinates to obtain the values of n_1 and n_2. The constant C_1 can now be evaluated by substituting the values of exponent $1/n$, $1/n_1$ and $1/n_2$ in equation (8.3). The number of times the tests need to be repeated depends upon the accuracy desired. In most practical situations, it is often necessary to check the effect of cutting speed for only a few feed rates and depths of cut.

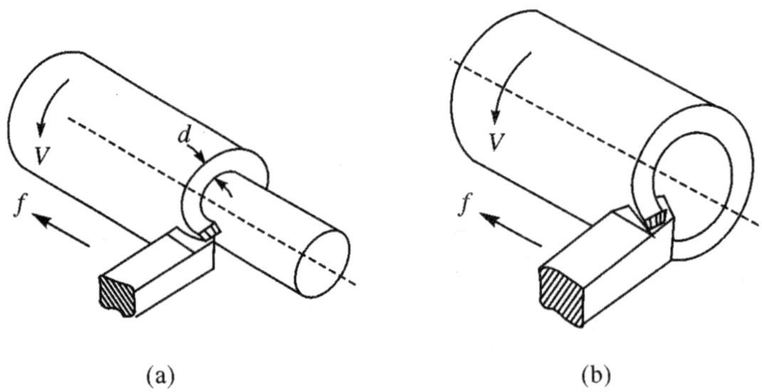

Fig. 8.11 Tool-life testing

8.7 QUICK TOOL-LIFE TESTING

The tool-life testing method described above is obviously time consuming and expensive. In order to reduce the testing time and cost, several quick tool-life testing techniques have been suggested. These include (a) facing tests, (b) taper turning tests and (c) disc turning tests. In facing tests, suggested by Solaja (1955) and Lorenz (1963-64), the tool traverses radially outwards from diameter D_1 to D_2 thereby changing the cutting speed continuously from V_1 to V_2. The effect of variable speed is evaluated in terms of an equivalent cutting speed V_{eq}. The *equivalent speed* is usually defined as the constant speed which would give the same tool wear in the same cutting time as the variable speed.

Assuming linear wear-land growth as approximated in Fig. 8.12,

$$\frac{dl_w}{dT} = \frac{\text{Flank wear land criterion}(l_w^*)}{\text{Tool life}(T)} \qquad ...(8.7)$$

If Taylor tool-life equation is assumed to hold in the range of cutting speed covered while the tool travels from diameter D_1 to D_2, the wear land l_{w1} at any time T_1 can be obtained as

$$l_{w1} = \int_0^{T_1} \frac{l_w^*}{C^{1/n}} V^{1/n} dT \qquad ...(8.8)$$

The corresponding equivalent speed V_{eq} can now be expressed as

$$V_{eq} = \left[\frac{1}{T_1}\int_0^{T_1} V^{1/n} dT\right]^n \qquad ...(8.9)$$

Since the equivalent speed V_{eq} will also give the same flank wear land l_{w1} after time T_1, therefore,

$$l_{w1} = \frac{l_w^*}{C^{1/n}} V_{eq}^{1/n} T_1 \qquad ...(8.10)$$

In facing tests the cutting speed varies with diameter and time as

$$V = 2\pi RN = 2\pi N(fNT + R_1) \qquad ...(8.11)$$

where N is the spindle speed, f is the feed per revolution and T is the cutting time to reach radius R.

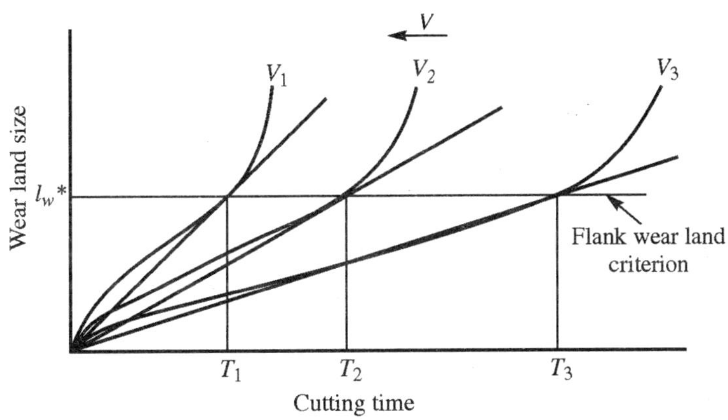

Fig. 8.12 Approximated linear wear land growth

From equations (8.9) and (8.11), the equivalent speed V_{eq} in facing from radius R_1 to R_2 in time T_1 can be obtained as

$$V_{eq} = 2\pi N\left[\frac{1}{T_1}\int_0^{T_1} (fNT + R_1)^{1/n} dT\right]^n \qquad ...(8.12)$$

or

$$V_{eq} = \pi N\left[\frac{n}{n+1}\left(\frac{D_2^{\frac{1+n}{n}} - D_1^{\frac{1+n}{n}}}{D_2 - D_1}\right)\right] \qquad ...(8.13)$$

since

$$T_1 = \frac{R_2 - R_1}{fN} = \frac{D_2 - D_1}{2fN} \qquad ...(8.14)$$

For a given D_1 and D_2, $V_{eq} \propto N$ and the effect of cutting speed on tool life can be obtained by varying N.

The fraction of tool life X consumed in time T_1 when tool moves from D_1 to D_2 is

$$X = \frac{1}{C^{1/n}} V_{eq}^{1/n} T_1 \quad ...(8.15)$$

or

$$X = \left(\frac{\pi N}{C}\right)^{1/n} \left[\frac{n}{n+1}\left(\frac{D_2^{\frac{1+n}{n}} - D_1^{\frac{1+n}{n}}}{2fN}\right)\right] \quad ...(8.16)$$

from equations (8.10), (8.13) and (8.14). If X_1 and X_2 are the fraction of tool life consumed at spindle spees N_1 and N_2 for a given D_1, D_2 and f, then

$$\frac{X_1}{X_2} = \left(\frac{N_1}{N_2}\right)^{\frac{1-n}{n}} \quad ...(8.17)$$

or

$$n = \frac{1}{[\log (X_1/X_2)/\log (N_1/N_2)] + 1} \quad ...(8.18)$$

and

$$C = \pi N \left[\frac{n}{n+1}\left(\frac{D_2^{\frac{1-n}{n}} - D_1^{\frac{1+n}{n}}}{2fNX}\right)\right]^n \quad ...(8.19)$$

Thus, by performing tests at two different speeds for a set of diameters D_1 and D_2, the tool life constants C and n can be evaluated.

In *taper turning tests*, suggested by Heginbotham and Pandey (1966), as the tool moves from diameter $D_1 = 2R_1$and $D_2 = 2R_2$(Fig. 8.13) there is continuous variation in cutting speed. The effect of speed is, therefore, evaluated in terms of equivalent cutting speed.

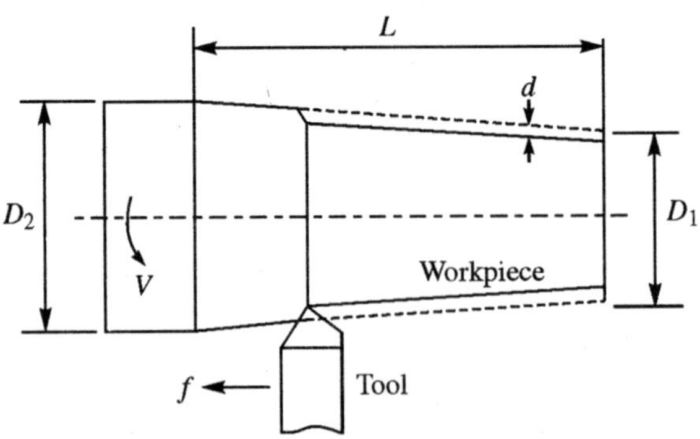

Fig. 8.13 Taper turning

In taper turning, the speed at any given radius R or diameter D (Fig. 8.13) is

$$V = 2\pi RN + 2\pi N\left[R_1 + (R_2 - R_1)\frac{Nf}{L}T\right] \quad ...(8.20)$$

and

$$V_{eq} = 2\pi RN\left[\frac{1}{T_1}\int_0^{T_1}\left\{R_1 + (R_2 - R_1)\frac{Nf}{L}T\right\}dT\right]^n \qquad \text{....(8.21)}$$

The fraction of tool life consumed,

$$X = \left(\frac{\pi N}{C}\right)^{1/n}\left(\frac{n}{n+1}\right)\left(\frac{L}{fN}\right)\left(\frac{R_2^{\frac{1+n}{n}} - R_1^{\frac{1+n}{n}}}{R_2 - R_1}\right) \qquad \text{...(8.22)}$$

By performing two taper turning tests at spindle speeds N_1 and N_2, the corresponding values of X_1 and X_2 can be obtained and n can be evaluated using equation (8.18). The value of C can be found by substituting for n in equation (8.22). Comparison between tool-life equation obtained from this test and that obtained by constant speed test shows good agreement (Fig. 8.14).

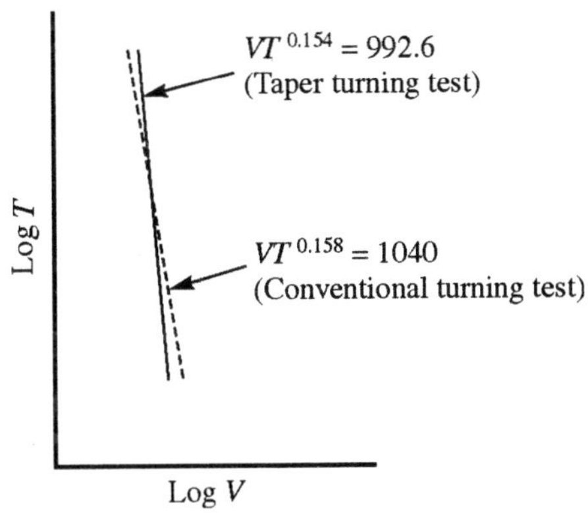

Fig. 8.14 Comparison of conventional and taper turning tool life curves (After Heginbotham and Pandey, 1966)

In *disc turning test* (Fig. 8.15) suggested by Rao and Lal (1977), the workpiece consisting of discs cut from a plate are fixed on a mandrel with spacers in between adjacent discs. The tool is fed into the workpiece using the infeed drive such that orthogonal cutting conditions are obtained. As the tool moves from outside diameter D_1 to a smaller diameter D_2, the continuously varying speed is evaluated in term of equivalent speed V_{eq}.

Proceeding in the same manner as for the facing test, the equivalent speed can be obtained as

$$V_{eq} = \pi N\left[\frac{n}{n+1}\left(\frac{D_1^{\frac{1+n}{n}} \quad D_2^{\frac{1+n}{n}}}{D_1 - D_2}\right)\right]^n \qquad \text{...(8.23)}$$

and fraction of tool life X as

$$X = \left(\frac{\pi N}{C}\right)^{1/n}\left[\frac{n}{n+1}\left(\frac{D_1^{\frac{1+n}{n}} - D_2^{\frac{1+n}{n}}}{2fN}\right)\right] \qquad \text{...(8.24)}$$

In these equations $D_1 > D_2$, being the outer diameter. In facing tests $D_2 > D_1$ since the tool moves outwards.

Fig. 8.15 Disc turning test (After Rao and Lal, 1977)

As before, if X_1 and X_2 are evaluated at two different spindle speeds N_1 and N_2 for a set of diameters D_1 and D_2 and infeed f, the tool-life exponent n can be obtained from equation (8.18). The tool-life constant C can be obtained by substituting for n in equation (8.24). Tool life results obtained by this method are shown in Fig. 8.16.

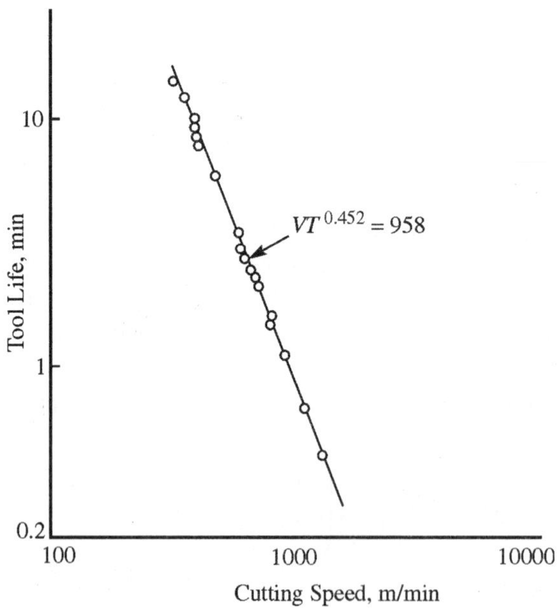

Fig. 8.16 Tool life-cutting speed relationship from disc test (After Rao and Lal, 1977)

Facing tests suggested by Saloja (1955) cannot be used for tungsten carbide tools, as either very high spindle speed or workpiece of very large diameter will be required to complete a run to failure in one pass. This test also suffers from the assumption that the material is homogeneous and has identical properties from outer to inner core. Taper and disc turning tests eliminate this drawback. Disc turning tests can be carried out at high cutting speeds for both flank and crater wear results.

8.8 MACHINABILITY

The term machinability is often used to describe the machining properties of work materials. A material is said to have good machinability if the tool wear is low, the material removal rate is high, the surface finish produced is good and the power consumption is low. The two other considerations are ease of chip disposal and dimensional accuracy. Clearly, these parameters will depend upon the cutting conditions, the tool geometry, the tool material, the properties of the workpiece material, etc. The material properties that affect machinability are shear strength, strain hardening, hardness, abrasiveness, thermal conductivity, chemical composition, structure, etc. Since the parameters that affect machinability are numerous, machinability appears to be a difficult concept to reduce to quantitative terms. Inspite of this, many attempts have been made to obtain a quantitative measure of machinability in terms of a 'Machinability Index' — a single, unique index to describe the relative ease with which a material could be machined. The criterion used in these attempts could be classified into the following four broad categories:

(*a*) Tool-life criterion
(*b*) Production-rate criterion
(*c*) Power-consumption criterion
(*d*) Surface-finish criterion

When tool-life criterion is used, it is common to express it in terms of cutting speed for a given tool life. For assessing machinability, a common material is chosen as a standard and the machinability of other materials is compared and expressed as a ratio. The material recommended as standard material is B1112 steel and its machinability index is taken as 1 or 100%. ISO standard provides machinability index numbers based on 30 min of tool life and 0.33 mm flank wear as the tool failure criterion (Fig. 8.17). For example, if $(V_T)s$ is the cutting speed for a tool life of T min for the standard material and $(V_T)_M$ is the cutting speed for the same tool life for the test material then the machinability of the test material can be expressed as $[(V_T)_M/(V_T)_S]\times 100\%$. Clearly, a material with a higher cutting speed for a given tool life will have better machinability. In this method, the effect of surface finish is generally not accounted. Further, if different tool materials are used to assess machinability, different ratings may occur.

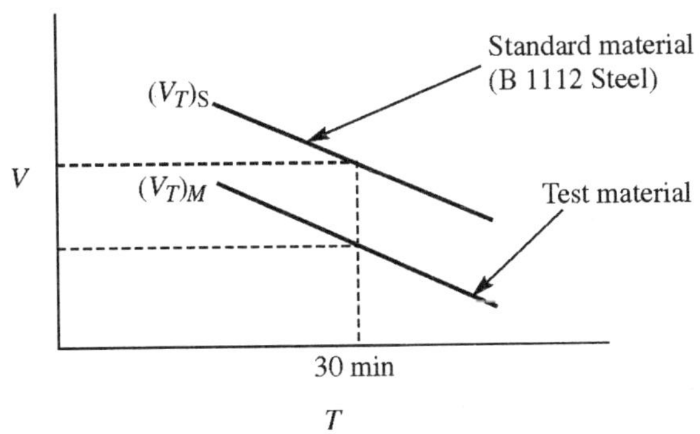

Fig. 8.17 Machinability rating

The tool life may also be expressed in terms of material removal rate and the production rate may be taken as the machinability criterion. Clearly, the higher the cutting speed for a given tool life, the greater will be the material removed and the higher will be the production rate. In this case, the surface finish could also be considered by altering the cutting conditions to obtain the best production rate for a fixed tool life and required surface finish. By comparing the best production rate of the test material with that of the standard, the machinability rating may be obtained.

Considering the rigidity of machine tools, it is necessary to limit the value of cutting forces to avoid excessive vibration and chatter during machining. The power consumption during machining is related to the cutting force—the higher the cutting force, the greater is the power consumption. When the cutting force is high a large machine tool may be required which in turn increases the production cost. Thus, materials requiring higher cutting forces will have lower machinability index.

The surface-finish requirements also vary widely. Some materials may permit use of higher cutting speed or induce lower cutting forces but give poor surface finish. In such situations, the surface-finish criterion is important. Thus, a material that produces better surface finish under a given set of conditions may be considered to have better machinability rating.

It is clear that any rating regarding machinability is valid only for the set of conditions used when the observations were made. For example, for a given set of cutting conditions, material M_1 may have higher machinability rating than material M_2 but under a different set of cutting conditions, material M_2 may turn out to be better than M_1. Similarly, a different set of ratings may be obtained when, say, a different tool material is used. Further the rating of materials from surface finish considerations may be different from the one obtained on the basis of tool life or power consumption. The term machinability is, therefore, meaningful only in a loose qualitatives sense and should only be used as a guide to machining properties.

CHAPTER 9

Economics of Machining

9.1 INTRODUCTION

Machining operations are usually judged in terms of tool life, cutting forces, surface roughness, dimensional accuracy and material removal rate. The objective is to produce a component of required dimensions and surface finish at the minimum possible cost. The factors that can be varied during a machining operation to change the cost are: speed, feed, depth of cut, tool and work material, tool geometry and cutting fluid. It is clear that the variables that affect the machining cost are numerous and it is difficult to examine their effects simultaneously. Factors such as tool and work material, tool geometry and cutting fluid can be selected from other considerations and information regarding these are available in handbooks and manufacturer's guide–books. Technical data regarding cutting conditions (speed, feed and depth of cut) are usually not available and their optimum selection cannot be made from cost considerations only.

The total cost of producing a component involves other factors; the two main factors being the cost of material and the cost of material handling. In general, these two factors, form a substantial part of the total manufacturing cost. It is not proposed to discuss these aspects here or to discuss the economics of different machine tools, but simply to concentrate on those costs which are involved when the component is actually put on the machine tool. Thus, the nature and effects of cutting conditions on machining cost must be examined. For example, when the cutting speed is low, longer tool life is obtained but the production rate is low. On the other hand, at high cutting speed although production rate increases, the tool life is very short. This would again make the operation uneconomical since the cost of tool regrinding and replacement will be high. Clearly, an optimum cutting speed exists for each application. This would mean a compromise between production rate and tool cost.

Similarly, higher values of feed and depth of cut would decrease the machining time but would adversely affect the tool cost. Thus, the cost of machining is minimum at a particular set of cutting conditions (speed, feed and depth of cut). For simplicity, we will only consider the case of a single pass straight turning operation and obtain the value of optimum cutting speed.

9.2 MACHINING COST

The machining cost can be classified as:

(*a*) Non-productive cost

(*b*) Cutting cost

(*c*) Tool cost

9.2.1 Non-productive Cost

Non-productive cost can be evaluated by adding all non-productive time and multiplying it by the cost rate. The relevant non-productive times are:

1. Initial set-up time: T_s min. (Occurring once per batch)
2. Loading and unloading time: T_l min. (Occurring once per piece)
3. Tool advance and withdrawal time: T_a min. (Occurring once per piece)
4. Tool removing and replacing time : T_r min. (Occurring once per tool regrind)
5. Idle time: T_i min.

Assuming there are N_B number of components in a batch and N_g number of components are produced between each tool regrinds, then the total non-productive time T_N will be

$$T_N = T_s + (T_l + T_a)N_B + T_r\left(\frac{N_B}{N_g}\right) + T_i \quad ...(9.1)$$

The cost rate c_R will include the labour cost rate c_L, the overhead cost rate c_O and the depreciation cost rate c_D. Thus,

$$c_R = (c_L + c_O + c_D) \quad ...(9.2)$$

and the non-productive cost C_N will be

$$C_N = c_R\, T_N \quad ...(9.3)$$

and

$$C_N = (c_L + c_O + c_D)\left[T_s + (T_l + T_a)N_B + T_r\left(\frac{N_B}{N_g}\right) + T_i\right] \quad ...(9.4)$$

Here and in all subsequent discussions, the cost rates are in units of cost per min and the relevant times are in minutes.

9.2.2 Cutting Cost

Cutting cost C_C can be evaluated by multiplying total cutting time with cost rate. Thus,

$$C_C = (c_L + c_O + c_D)T_c N_B \quad ...(9.5)$$

where T_c is the cutting time per component.

9.2.3 Tool Cost

Tool cost C_T will include the initial cost of tool C_i and the cost of tool regrinding C_g. Thus,

$$C_T = \left(\frac{C_i}{r_g} + C_g\right)\frac{N_B}{Ng} \quad ...(9.6)$$

where r_g is the number of regrinds possible on a tool.

The machining cost per batch of components excluding the material cost, therefore, becomes

$$C_B = C_N + C_C + C_T \quad ...(9.7)$$

or

$$C_B = (c_L + c_O + c_D)[T_s + (T_l + T_a)N_B + T_r\left(\frac{N_B}{N_g}\right) + T_i + T_c\, N_B] + \left(\frac{C_i}{r_g} + C_g\right)\frac{N_B}{N_g} \quad ...(9.8)$$

and the machining cost per piece C_P is

$$C_P = C_B/N_B \qquad ...(9.9)$$

It is clear that the machining cost can be reduced by decreasing the non-productive time. In this regard, the use of jigs and fixtures, improved tool holder design, etc. could be very effective. Use of appropriate tool material and tool geometry would give longer tool life, thus reducing the number of tool grinding and replacements. It is generally not possible to reduce the overhead cost as long as appropriate machine tool is in use. The other factor that can reduce the total cost is the cutting cost. Increasing the cutting speed as well as feed will reduce the cutting time but would adversely affect the tool life and increase the tool grinding and tool removing and replacing cost. Therefore, we must use optimum cutting conditions and a variety of criteria have been used for optimization. These include

(*a*) minimum cost per component,

(*b*) maximum production rate, and

(*c*) maximum profit rate criteria.

While minimum cost per piece has merits but it is not necessarily the best criterion. Minimizing cost per piece maximizes profit per piece but the profit made over a period of time is not optimum since both the production cost and production time govern the profit made in a given time. From the manufacturer's point of view, the maximization of profit rate without increasing the price of the commodity may be of greater interest.

The time T_B required for machining of a batch of components is

$$T_B = T_s + (T_l + T_a)N_B + T_r\left(\frac{N_B}{N_g}\right) + T_i + T_c N_B \qquad ...(9.10)$$

and the number of components machined between each tool grind is

$$N_g = \frac{T}{T_c} \qquad ...(9.11)$$

where T is the tool life. Thus,

$$T_B = T_s + (T_l + T_a)N_B + T_r\left(\frac{T_c}{T}\right)N_B + T_i + T_c N_B \qquad ...(9.12)$$

and the machining time T_P per piece is

$$T_P = \frac{1}{N_B}(T_s + T_i) + \left(T_l + T_a + T_c + T_r\frac{T_c}{T}\right) \qquad ...(9.13)$$

The inverse of T_P gives the production rate R_P, *i.e.*, number of pieces produced per unit time. Thus,

$$R_P = \frac{1}{\frac{1}{N_B}(T_s + T_i) + \left(T_l + T_a + T_c + T_r\frac{T_c}{T}\right)} \qquad ...(9.14)$$

Clearly the production rate can be increased by reducing the non-productive time. Again optimum values of cutting speed and feed should be chosen since these and tool life have opposing effects on the production rate.

The profit rate P_R can be expressed as

$$P_R = \frac{I_p - C_p}{T_p} \qquad ...(9.15)$$

where I_p is the income per piece excluding material cost and C_P and T_P can be evaluated using equations (9.9) and (9.13). It is clear that increased profit rate can be obtained by reducing the cost per component and by increasing the production rate. Since variations in cutting conditions give optimum values of cost per component as well as production rate, a maximum profit rate will also occur.

9.3 OPTIMUM CUTTING SPEED

To find the optimum cutting speed for a batch of components, it is necessary to know the tool life-cutting speed relationship. It has already been shown that an empirical relationship exists between these two and it is of the form

$$VT^n = C \qquad ...(9.16)$$

or

$$T = \left(\frac{C}{V}\right)^{1/n} \qquad ...(9.17)$$

where V is the cutting speed and n and C are constants for a particular tool and workpiece material and for a particular set of values for feed f and depth of cut d. Again for simplicity, we will only consider the case of a single pass straight turning operation where feed and depth of cut can be fixed from the considerations of surface finish, maximum power of motor and maximum cutting force that the machine can bear without affecting the dimensional accuracy. As already been discussed, the cutting speed has greatest influence on tool life, followed by feed and depth of cut. Thus, an increase in depth of cut will cause smallest decrease in tool life, followed by feed and cutting speed. In other words, when an increase in production rate is required in rough machining, the depth of cut and then the feed should be set at the maximum possible values before choosing the optimum cutting speed.

The cutting time T_c for a cylindrical workpiece of length L and diameter D for a single pass operation is given by

$$T_c = \frac{\pi DL}{fV} \qquad ...(9.18)$$

or

$$T_c = \frac{K_1}{V} \qquad ...(9.19)$$

where

$$K_1 = \frac{\pi DL}{f} \qquad ...(9.20)$$

9.3.1 Cutting Speed for Minimum Cost

The machining cost per batch of components excluding the material cost is given by equation (9.8). Using equations (9.11), (9.17) and (9.19), equation (9.8) can be rewritten as

$$CB = (c_L + c_O + c_D)\left[T_s + (T_l + T_a)N_B + T_r\left(\frac{T_c}{T}\right)N_B + T_i + \frac{K_1}{V}N_B\right] + \left(\frac{C_i}{r_g} + C_g\right)\left(\frac{T_c}{T}\right)N_B \qquad ...(9.21)$$

or

$$C_B = (c_L + c_O + c_D)[N_B(T_l + T_a) + T_s + T_i] + \frac{N_B(V)^{\frac{1-n}{n}}}{K_2}$$

$$\left[(c_L + c_O + c_D)T_r + \frac{C_i}{r_g} + C_g\right] + (c_L + c_O + c_D)N_B K_1 V^{-1} \quad ...(9.22)$$

since

$$N_g = \frac{T}{T_c} = \frac{V}{K_1}\left(\frac{C}{V}\right)^{1/n} = K_2(V)^{\frac{n-1}{n}} \quad ...(9.23)$$

where

$$K_2 = \frac{C^{1/n}}{K_1} \quad ...(9.24)$$

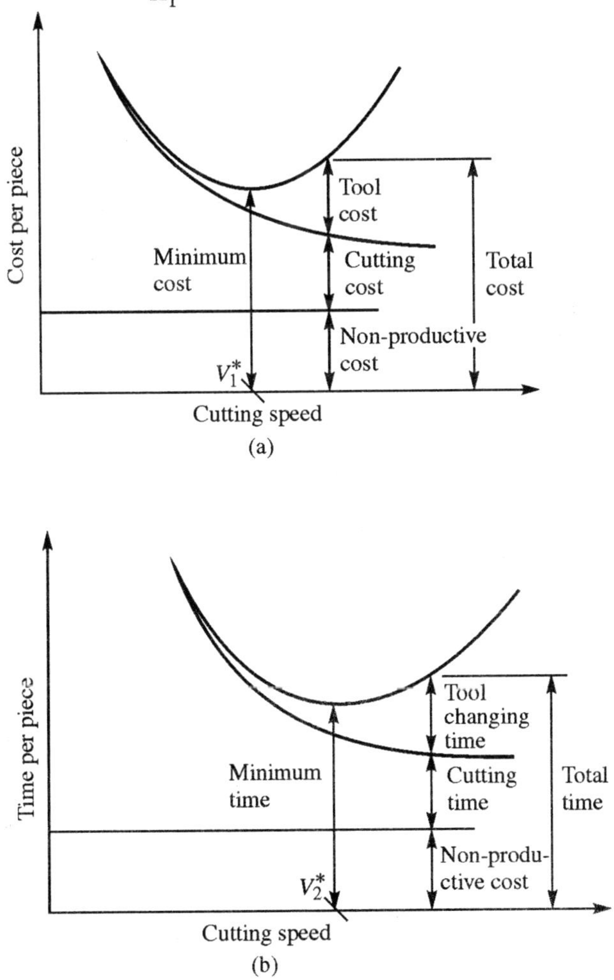

Fig. 9.1 Variation of cost and time per piece with cutting speed

The variation of cost and time per piece with cutting speed is shown in Fig. 9.1 where the effect of cutting speed on various cost and time components can be clearly seen. This shows that the non-productive cost and non-productive time remain constant, the tool cost and tool changing time increase with increase in cutting speed and the cutting cost and cutting time decrease with increase in cutting speed. These factors give optimum values for cutting speed for minimum cost and minimum machining time or maximum production rate.

For obtaining the optimum cutting speed V_1^* for minimum cost per batch of component, equation (9.22) must be differentiated with respect to V and equated to zero. Thus,

$$\frac{dC_B}{dV} = \frac{(1-n)N_B}{nK_2}\left[(c_L + c_O + c_D)T_r + \frac{C_i}{r_g} + C_g\right](V)^{\frac{1-2n}{n}}$$

$$-(c_L + c_O + c_D)N_B K_1 V^{-2} = 0 \qquad \text{...(9.25)}$$

and the optimum cutting speed is obtained as

$$V_1^* = \left[\frac{nK_1 K_2 (c_L + c_O + c_D)}{(1-n)(c_L + c_O + c_D)T_r + \frac{C_i}{r_g} + C_g}\right]^n \qquad \text{...(9.26)}$$

9.3.2 Cutting Speed for Maximum Production Rate

The production rate per minute R_P is given by equation (9.14). Using equations (9.19) and (9.23), R_P is obtained as

$$R_p = \frac{1}{\frac{l}{N_B}(T_s + T_i) + \left(T_l + T_a + K_l V^{-1} + T_r \frac{(V)^{\frac{1-n}{n}}}{K_2}\right)^2} \qquad \text{...(9.27)}$$

For obtaining the optimum cutting speed V_2^* for maximum production rate for a batch of components, equation (9.27) must be differentiated with respect to V and equated to zero. Thus,

$$\frac{dR_p}{dV} = \frac{-K_1 V^{-2} + \frac{T_r}{K_2}\left(\frac{1-n}{n}\right)(V)^{\frac{1-2n}{n}}}{\left[\frac{1}{N_B}(T_s + T_i) + \left(T_l + T_a + K_1 V^{-1} + \frac{T_r}{K_2}\right)(V)^{\frac{1-n}{n}}\right]^2} = 0 \qquad \text{...(9.28)}$$

giving the optimum speed as

$$V_2^* = \left[\frac{nK_1K_2}{(1-n)T_r}\right]^n \qquad ...(9.29)$$

9.3.3 Cutting Speed for Maximum Profit Rate

The profit rate P_R is given by equation (9.15). Using equations (9.8), (9.9), (9.17) and (9.19), the cost per piece C_P can be expressed as

$$C_P = (c_L + c_O + c_D)[T_l + T_a + \frac{1}{N_B}(T_s + T_i)] + \frac{(V)^{\frac{l-n}{n}}}{K_2}$$

$$[(c_L + c_O + c_D)T_r + \frac{C_i}{r_g} + C_g] + (c_L + c_O + c_D)K_1 V^{-1} \qquad ...(9.30)$$

The machining time T_P per piece is given by the equation (9.13). Using equations (9.17) and (9.19), T_P can be expressed as

$$T_P = \frac{1}{N_B}(T_s + T_i)\left(T_i + T_a + K_1V^{-1} + \frac{T_r}{K_2}(V)^{\frac{1-n}{n}}\right) \qquad ...(9.31)$$

The profit rate is therefore,

$$P_R = \frac{I_P - [(c_L + c_O + c_D)\left\{T_l + T_a + \frac{1}{N_B}(T_s + T_i)\right\} + \frac{(V)^{\frac{1-n}{n}}}{K_2}\left\{(c_L + c_O + c_D)T_r + \frac{C_i}{r_g} + C_g\right\} + (c_L + c_O + c_D)K_1V^{-1}]}{\frac{1}{N_B}(T_s + T_i)\left(T_l + T_a + K_1V^{-1} + \frac{T_r}{K_2}(V)^{\frac{1-n}{n}}\right)} \qquad ...(9.32)$$

or

$$P_R = \frac{I_p - \left(\frac{C_i}{r_g} + C_g\right)\frac{(V)^{\frac{1-n}{n}}}{K_2}}{\left[T_l + T_a + \frac{1}{N_B}(T_s + T_i)K_1V^{-1} + \frac{T_r}{K_2}(V)^{\frac{1-n}{n}}\right]} - (c_L + c_O + c_D) \qquad ...(9.33)$$

The cutting speed for maximum profit rate can now be obtained from

$$\frac{dP_R}{dV} = 0 \qquad ...(9.34)$$

In order to simplify the analysis, the cutting conditions are generally optimized without considering any increases in the income per piece or the selling price of the commodity. Thus, I_P is a constant and equations (9.33) and (9.34) on simplification gives,

$$\left(\frac{1-n}{n}\right)\frac{T_r}{K_2}\left[I_p - 2\left(\frac{C_i}{r_g}+C_g\right)\right](V_3^*)^{\frac{1-n}{n}} - K_1\left[I_p+\frac{1}{n}\left(\frac{C_i}{r_g}+C_g\right)\right](V_3^*)^{-1}$$

$$+\left(\frac{1-n}{n}\right)\left(\frac{C_i}{r_g}+C_g\right)\left[T_1+T_a+\frac{1}{N_B}(T_s+T_i)\right]=0 \qquad ...(9.35)$$

The solution of this equation gives the value of V_3^*, the cutting speed for maximum profit rate.

9.4 RESTRICTIONS ON CUTTING CONDITIONS

The final choice of the optimum value of cutting speed will have to satisfy a number of restrictions. These are given below.

(*a*) Maximum Power Restriction

The power required during a cutting operation (P_w) is a function of the cutting condition and may be expressed as

$$P_w = B_w V f^{m_1} d^{m_2} \qquad ...(9.36)$$

or

$$P_w = B'_w V \qquad ...(9.37)$$

where

$$B'_w = B_w f^{m_1} d^{m_2} = \text{Constant}$$

m_1 and m_2 are also constants for a given tool-work combination. The maximum power available on a machine tool is limited by the power capacity of the main drive. If the optimum speed is such that the available power is exceeded, the speed must be adjusted to meet this restriction. The reduction in speed should be such that maximum available power is used. This would ofcourse mean a change away from the minimum cost or maximum production rate or maximum profit rate.

(*b*) Speed Restriction

Most machine tools have cutting speeds available in steps over a certain range. On such machines, the step closest to the optimum value should be used. If the optimum speed found is beyond the range, the cutting speed must not exceed the upper limit. The lower limit of the cutting speed is generally limited by the formation of built-up edge.

(*c*) Force and Vibration Restriction

Machine components are designed for a maximum permissible load beyond which the tool-work deflections are excessive resulting in dimensional inaccuracies. The cutting force F_P can be expressed as

$$F_P = B_p f^{m_1} d^{m_2}$$

where B_P is a constant. Cutting speed has very little effect on cutting force and for all practical purposes its effect can be neglected. Thus, there is no restriction on cutting speed in terms of cutting force but this will limit the maximum values of feed and depth of cut that can be used. Excessive vibrations and chatter also puts restrictions on cutting conditions.

(d) Surface Finish Restrictions

The surface finish during a cutting operation depends upon the tool-work material, tool geometry, process geometry, cutting conditions and the type of coolant used. For a given operation, the surface finish restriction may be expressed in terms of cutting conditions as

$$h_m \geq B_s f^{a_1} V^{a_2}$$

where h_m is the maximum permissible surface roughness, and B_s, a_1 and a_2 are constants. From the point of view of surface finish, feed marks are the predominant phenomenon and this puts a serious restriction on *f*. Cutting speed affects the surface finish due to the formation of built-up-edge which gets smaller and smaller and disappears at sufficiently high speed, improving the surface finish.

9.5 COMPARISON OF THE THREE CRITERIA

A comparison of the three optimization criteria considered here is of considerable practical interest. It appears that maximum production rate criterion is the simplest one to use since no information regarding various cost rates is required. The minimum cost criterion is slightly more complex while the maximum profit rate criterion involves solution of complex equation using numerical technique. Each of these criteria yield different values for optimum cutting speed. The optimum cutting speed for minimum cost (equation 9.26) is always less than that for maximum production rate (equation 9.29). This is clearly seen by equating $\left(\frac{C_i}{r_g} + C_g\right)$ equal to zero in equation (9.26). These two optimum cutting speeds, *i.e.*, V_1^* and V_2^* may not maximise the profit rate which depends on the margin between the selling price and the cost of production as well as the rate of production. At the optimum value V_1^*, the production rate may be low while at speed V_2^* the cost of production may be high. Both of these conditions may lead to low profit margin. The optimum cutting speed V_3^* for maximum profit rate will, therefore, be different and will lie between V_1^* and V_2^*. When the optimum speeds are adjusted to take into account the limited number of speed steps and other restrictions, the final value of V_1^*, V_2^*, and V_3^* may not be widely different and it is often suggested that cutting conditions should be selected between minimum cost per component and maximum production rate.

CHAPTER 10

Practical Machining Operations

10.1 INTRODUCTION

Most practical machining operations use tools having two or more cutting edges but a close examination of these operations will reveal marked similarities with orthogonal and oblique cutting. The mechanics involved in some of the practical machining operations are, however, quite complex and in many cases only empirical relationships have been developed. The theoretical analyses that have been proposed are beyond the scope of this book but an attempt will be made to correlate the basic mechanics of the process with the process parameters. The influence of process variables on chip geometry, cutting forces and surface finish will also be considered.

10.2 TURNING

In straight turning operation the tool is made to travel parallel to the axis of rotation of the workpiece as shown in Fig. 10.1. The geometry of single-point tool used in this operation has already been discussed in Chapter 3. The cutting speed or the tangential cutting velocity V and the feed velocity v in this case are given by

$$V = \pi D_w N_w$$

and

$$v = fN_w \qquad ...(10.1)$$

where N_w is the rpm, D_w is the workpiece diameter and f is the feed rate. Since $V >> v$, it is reasonable to ignore the effect of feed velocity while considering the resultant relative velocity between the tool and the workpiece. Also, the depth of cut d is much smaller than D_w, therefore, the cutting speed may be assumed to be constant throughout the depth of cut.

10.2.1 Undeformed Chip-thickness

In turning operations, although the conditions of orthogonal cutting are not fully satisfied but for the purpose of application of the theory of orthogonal cutting discussed in chapter 4 we shall assume it to be the case when the principal cutting edge of the tool is perpendicular to the cutting velocity vector, *i.e.*, $i = 0$ in ORS and NRS systems. In turning, the material removal mainly takes place along the principal cutting edge *ab* (Fig. 10.1) but some material removal also occurs along the auxiliary cutting edge *bc* and because of this the operation is often called restricted orthogonal cutting. For tool geometry specified in ASA and MRS systems, the orthogonality can be checked using equations (3.4) and (3.11).

The undeformed chip-thickness t and the underformed chip-width b, which define the geometry of the underformed chip in straight-turning operation, can be obtained as

$$t = f \cos \gamma_s$$

and

$$b = d / \cos \gamma_s \quad \text{...(10.2)}$$

where γ_s is the side cutting-edge angle in ASA system. In terms of the principle cutting-edge angle γ_p of ORS, MRS and NRS systems, $\gamma_s = (90 - \gamma_p)$. The rake angle α for equating this process to orthogonal cutting should be α_n in NRS system and i should be equal to zero. α_n can be evaluated in terms of other tool angles using equations (3.5), (3.9) and (3.12). When $i = 0$, α_n is equal to α_o of ORS system.

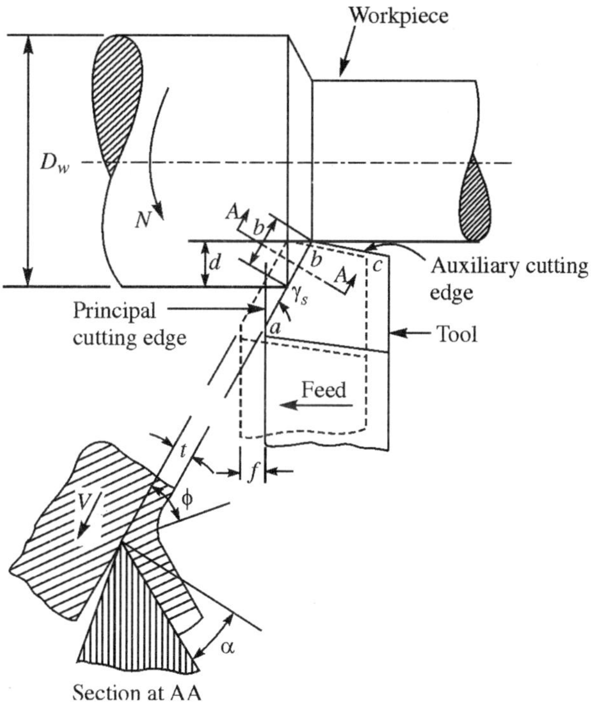

Fig. 10.1 Undeformed chip thickness in straight turning

10.2.2 Forces and Power

Most conventional turning operations approximate oblique cutting and require three components to specify the forces. The three components are usually specified as the cutting force F_p in the direction of cutting velocity, the feed force F_q in the direction of feed motion and the thrust force F_r perpendicular to the surface generated as shown in Fig. 10.2. These directions are chosen because of ease of measurement with the help of a three-component tool dynamometer such as one shown in Fig. 6.10. The force system gives the resultant force R as

$$R = \sqrt{(F_p^2 + F_q^2 + F_r^2)} \quad \text{...(10.3)}$$

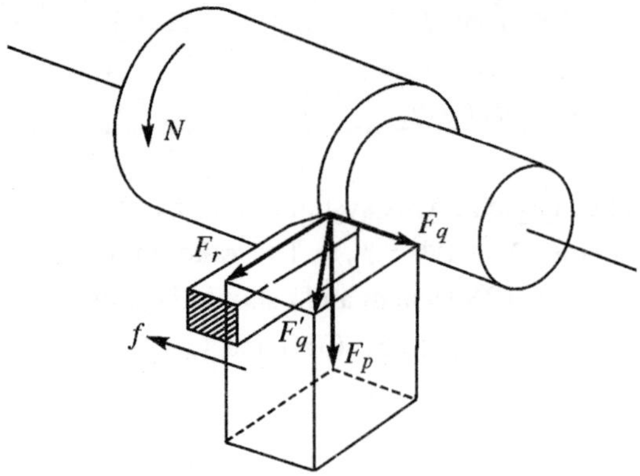

Fig. 10.2 Force components during straight turning

The three-dimensional force system can be reduced to a two-dimensional force system if the forces are considered in the orthogonal plane, *i.e.*, in a plane perpendicular to the principal cutting-edge with F'_q acting in the direction perpendicular to the principal cutting-edge in the horizontal plane. Thus,

$$R = \sqrt{(F_p^2 + F'^2_q)} \qquad ...(10.4)$$

and

$$F'_q = \sqrt{(F_q^2 + F_r^2)} \qquad ...(10.5)$$

This, however, is possible only when the inclination angle $i = 0$, $0 < \gamma_p < 90°$ and F'_q is contained in the orthogonal plane.

When $i = 0$ and $\gamma_p = 90°$ or $\gamma_s = 0$, the thrust force $F_r = 0$ and the two-dimensional force system gives

$$R = \sqrt{(F_p^2 + F_q^2)} \qquad ...(10.6)$$

Similarly, when $i = 0$ and $\gamma_p = 0$ or $\gamma_s = 90°$, the feed force $F_q = 0$ and the two-dimensional force system gives

$$R = \sqrt{(F_p^2 + F_r^2)} \qquad ...(10.7)$$

In the above turning cases, the auxiliary or end cutting-edge tends to deviate the chip-flow direction from the orthogonal plane and this in turn deviates the forces from the orthogonal plane. When the depth of cut d is large compared to the feed rate f, the deviation of forces from the orthogonal plane is small and can be neglected. In such situations where the turning operation is reduced to two-dimensional cutting (restricted orthogonal cutting), the theory developed in Chapter 4 can be applied for estimating the forces and other parameters. The value of α and t are to be the used as indicated in Fig. 10.1.

Force measurements carried out during turning of a variety of work materials with tools of various geometry and materials indicate that the forces are power functions of feed and depth of cut. Thus,

$$F_p = A_p f^{x_p} d^{y_p}$$

$$F_q = A_q f^{x_q} d^{y_q} \quad ...(10.8)$$

and

$$F_r = A_r f^{x_r} d^{y_r}$$

where parameters A, x and y with subscripts referring to the force components are constants for a given tool-workpiece combination and can be evaluated experimentaly. This empirical approach is, however, restrictive since the constants are unrelated to the fundamental and physical quantities.

The power required in turning P_w can now be evaluated from

$$P_w = F_p \bar{V} \quad ...(10.9)$$

where $\bar{V}$ is the mean cutting speed.

The specific cutting energy u_s defined as the energy required to remove a unit volume of material can be estimated from

$$u_s = \frac{P_w}{Z_m} \quad ...(10.10)$$

where Z_m, the material removal rate, is the product of the mean cutting speed $\bar{V}$ and the chip cross-sectional area A_c. Thus,

$$Z_m = \bar{V} A_c = \pi \bar{D} N_w f d \quad ...(10.11)$$

where $\bar{D}$ is the mean workpiece diameter. Using equations (10.9), (10.10) and (10.11), we get

$$u_s = \frac{F_p}{f d} \quad ...(10.12)$$

10.2.3 Surface Finish

The surface finish in machining operations depends on the type of chip formation, tool profile and geometry, cutting conditions etc., A continuous chip without built-up-edge represents a steady cutting condition and hence gives best surface finish. Built-up-edge formation leads to vibration and has adverse effect on surface finish. Discontinuous chip formation causes fluctuation in force and leads to the formation of ridges on the machined surface.

During continuous chip formation, the tool profile will be reproduced on the work surface in the form of feed marks. Figure 10.3 shows how the tool profile is reproduced on the work surface during straight turning with a sharp tool and a tool with finite nose radius.

The surface finish may be specified in terms of peak-to-valley height, centre-line average (CLA) or root-mean-square (RMS). The peak-to-valley height is the root-to-crest value of surface roughness as shown in Fig. 10.4.

The centre-line average value is found by averaging the height of the surface above and below a centre line over the sample length l_s. The centre line is a line parallel to the general profile direction such that the profile areas below and above the centre line are equal. If the ordinates y are y_1, y_2,...yn, then

$$h_{CLA} = \frac{1}{n}\sum_{i=1}^{n} |y_i| \quad ...(10.13)$$

or

$$h_{CLA} = \frac{1}{l_s}\int_{x=0}^{x=l_s} |y| dx \quad ...(10.14)$$

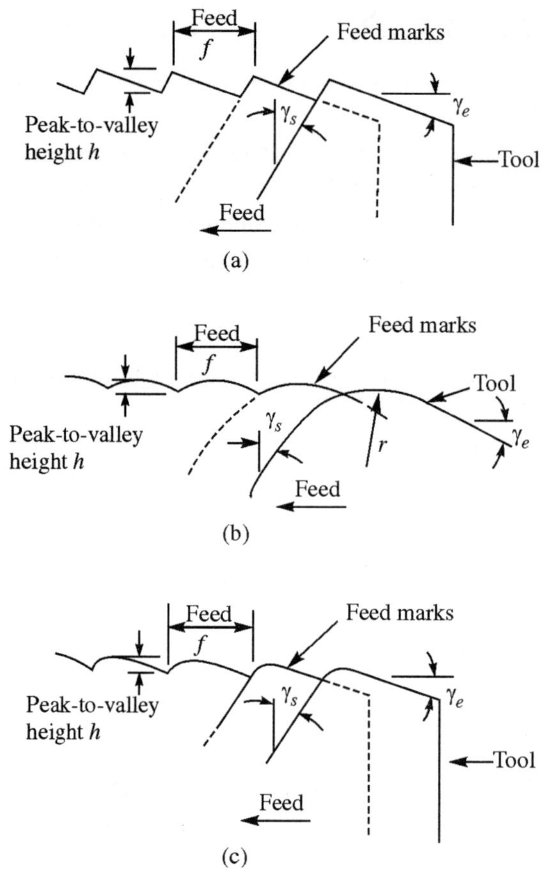

Fig. 10.3 Surface finish in turning (a) Sharp tool, (b and c) Tool with nose radius

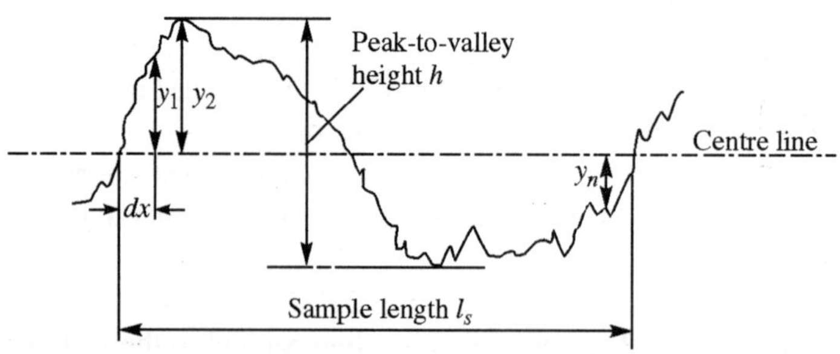

Fig. 10.4 Surface roughness profile

The RMS value is defined as the square root of the mean of the squares of the ordinate of the surface from the centre line. Thus,

$$h_{RMS} = \left[\frac{y_1^2 + y_2^2 + ...y_n^2}{n}\right]^{1/2} \quad ...(10.15)$$

or

$$h_{RMS} = \left[\frac{1}{l_s}\int_{x=0}^{x=l_s} y^2 dx\right]^{1/2} \quad ...(10.16)$$

Referring to Fig. 10.3(a), the peak-to-valley height h can be evaluated from

$$f = h\tan\gamma_s + h\cot\gamma_e \quad ...(10.17)$$

or

$$h = \frac{f}{\tan\gamma_s + \cot\gamma_e} \quad ...(10.18)$$

where γ_s and γ_e are the side and end cutting-edge angles in ASA system.

For a tool of finite nose radius r, Fig. 10.3(b), the peak-to-valley roughness can be evaluated from

$$r^2 - (r-h)^2 = \frac{f^2}{4} \quad ...(10.19)$$

or

$$h \approx \frac{f^2}{8r} \quad ...(10.20)$$

This expression is valid when the tool cutting is entirely on the nose radius only. When this is not so (Fig. 10.3c), the peak-to-valley roughness can be shown to be

$$h = f\tan\gamma_e + \frac{r}{2}\tan^2\gamma_e - \sqrt{(2fr\tan^3\gamma_e)} \quad ...(10.21)$$

It is clear that nose radius, cutting-edge angles and feed rate have significant influence on the surface finish produced. Effect of feed rate on surface finish evaluated experimentally seems to agree with equation (10.20). The general pattern of variation is shown in Fig. 10.5. The effect of cutting speed on surface finish is shown in Fig. 10.6 which indicates that better surface finish is realised only at higher cutting speeds where continuous chips without built-up-edge are formed. Increase in cutting speed beyond a certain values does not seem to have any significant effect on surface finish. At lower cutting speeds, discontinuous chip and built-up-edge formation results in poor surface finish.

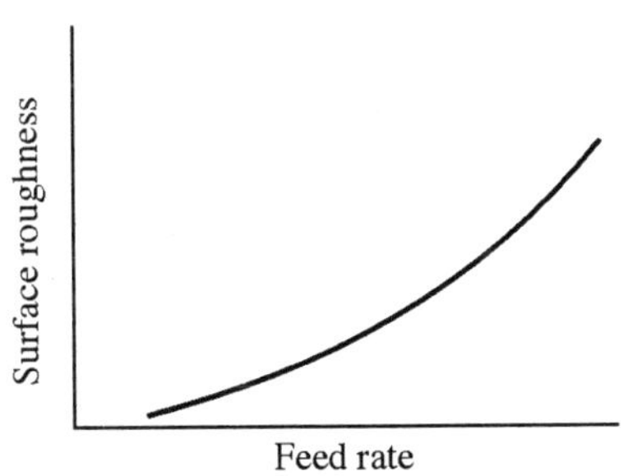

Fig. 10.5 Variation in surface roughness with feed

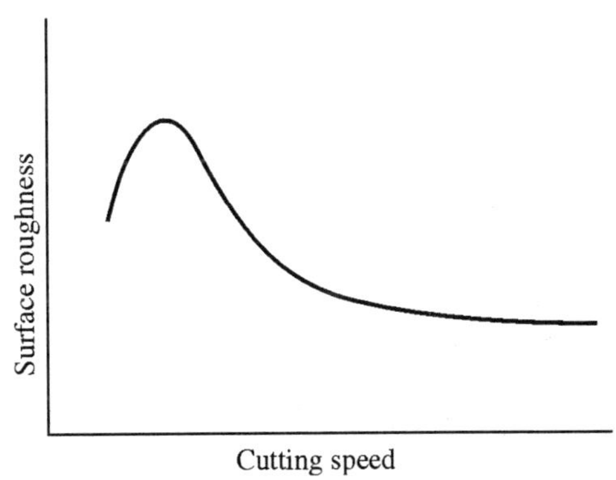

Fig. 10.6 Variation of surface roughness with cutting speed

10.3 SHAPING AND PLANING

The cutting action in shaping and planing is very similar to turning. In shaping, the cutting speed is provided to the tool and the feed is given to the workpiece, while in planing the feed is given to the tool and the workpiece is provided the cutting speed. In both cases, cutting takes place during the forward stroke only and the return speed is made as high as possible to reduce the overall machining time. The geometry of the cutting tools used and the cutting action are very similar to those for turning although the tool angles, feed and speed are not necessarily the same. Like in turning, the cutting is mainly performed along the principal cutting–edge *ab* while some material removal also takes place along the auxiliary cutting edge *bc* as shown in Fig. 10.7. In this case also the cutting at best can be called restricted orthogonal cutting when the inclination angle *i* is zero. We shall, however, assume this also to be a case of orthogonal cutting for the purpose of application of the orthogonal cutting theory.

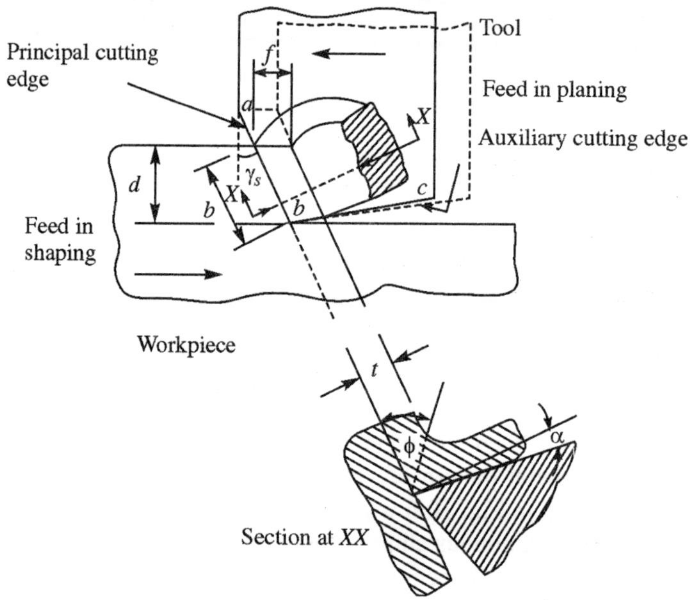

Fig. 10.7 Undeformed chip-thickness in shaping/planing

10.3.1 Undeformed Chip-thickness

Referring to Fig. 10.7, the undeformed chip-thickness *t* and the chip width *b* can be obtained as

$$t = f\cos\gamma_s \qquad ...(10.22)$$

and

$$b = d/\cos\gamma_s \qquad ...(10.23)$$

The rake angle α for equating to orthogonal cutting should be α_n which is equal to α_o when $i = 0$.

10.3.2 Forces

The resultant force can be resolved into three force components; F_p in the direction of cutting velocity, F_q in the direction of feed motion and F_r in the direction perpendicular to the surface generated as shown in Fig. 10.8. Thus,

$$R = \sqrt{(F_p^2 + F_q^2 + F_r^2)} \qquad ...(10.24)$$

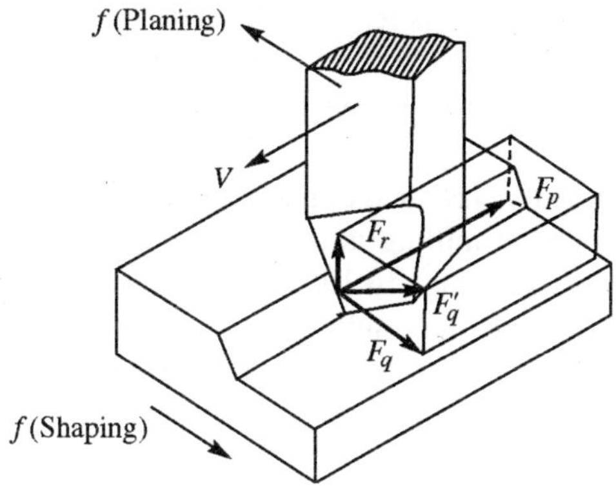

Fig. 10.8 Force components during shaping/planing

Considering a plane perpendicular to the principal cutting-edge, F_q and F_r component can be combined to give F'_q in the direction perpendicular to the principal cutting-edge. Thus,

$$R = \sqrt{(F_p^2 + F_q'^2)} \qquad ...(10.25)$$

This, however, is possible only when $i = 0$.

Equations for the cutting forces in turning may be applied for shaping and planing operations also but we must give due weightage for the intermittent nature of cuts while estimating the power requirement.

For estimating the force components using the orthogonal cutting theory developed in Chapter 4, equations (10.22) and (10.23) should be used for evaluating t and b, while $\alpha = \alpha_o$ when $i = 0$.

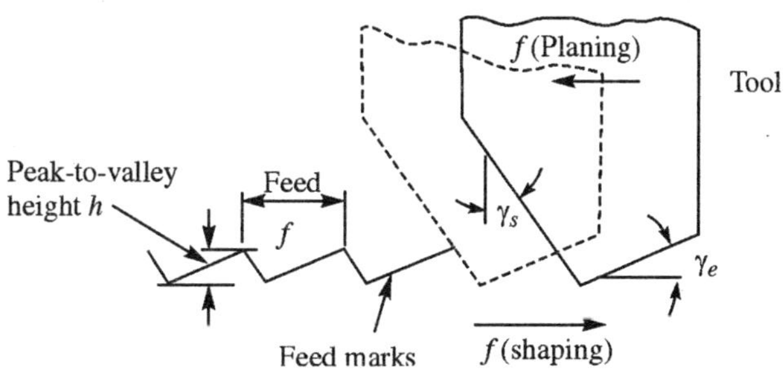

Fig. 10.9 Surface finish in shaping/planing

10.3.3 Surface Finish

The peak-to-valley height h for shaping/planing with a sharp tool (Fig. 10.9) can be obtained as

$$h = \frac{f}{\tan\gamma_s + \cot\gamma_e} \qquad ...(10.26)$$

which is the same as that for turning (equation 10.18). The results for the other types of tool can be obtained in a similar manner.

10.4 MILLING

In milling operations, material removal is achieved by a rotatory cutting tool with a number of cutting teeth rotating about a fixed axis. The workpiece is fed pass the cutter and each tooth removes material during interaction with the workpiece. The feed motion is, therefore, provided to the workpiece and the cutting speed to the cutter. The process provides intermittent cutting with each tooth producing a chip of variable thickness.

The milling operations can be broadly classified as plain (or peripheral) and face (or end) milling. When the cutter velocity during peripheral milling is in the opposite sense to the feed motion, the process is called *conventional* or *up-milling* (Fig. 10.10a) and when they are in the same sense (Fig. 10.10b), the operation is called *down-milling*. In up-milling the chip thickness is zero at the start of the cut and increases to a maximum value just before the tooth disengages the workpiece, while in down-milling the chip thickness is maximum at the start of cut and drops to zero value at the end of cut.

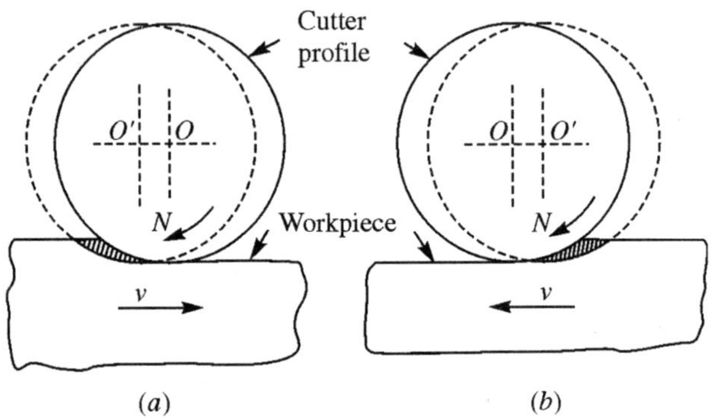

Fig. 10.10 Plain milling (a) Up-milling (b) Down-milling

In face milling, the cutter axis is perpendicular to the work surface and cutting is performed by cutting edge on the periphery and the face of the cutter.

10.4.1 Undeformed Chip-thickness

The cross-section of a chip produced in a typical plain milling operation is shown in Fig. 10.11. The path traced by the cutting edge is actually trochoidal but it has been approximated by a circular arc. This assumption is reasonably valid since the depth of cut d is usually much smaller than the cutter diameter D. The chip produced by a single cutting-edge will, therefore, be confined within the two arcs of radius equal to $D/2$ as shown.

Referring to Fig. 10.11, the chip length l can be evaluated by taking the chordal length ac. From triangles obc and abc, we get

$$l^2 = \left[\left(\frac{D}{2}\right)^2 - \left(\frac{D}{2} - d\right)^2\right] + d^2$$

or

$$l \approx \sqrt{dD} \qquad \text{...(10.27)}$$

If $\bar{t}$ is the mean chip thickness, then the volume of an individual chip can be evaluated as

$$(\text{Vol})_{\text{chip}} = l\bar{t}\,b = \bar{t}\,b\sqrt{dD} \qquad \text{...(10.28)}$$

where b is the chip width.

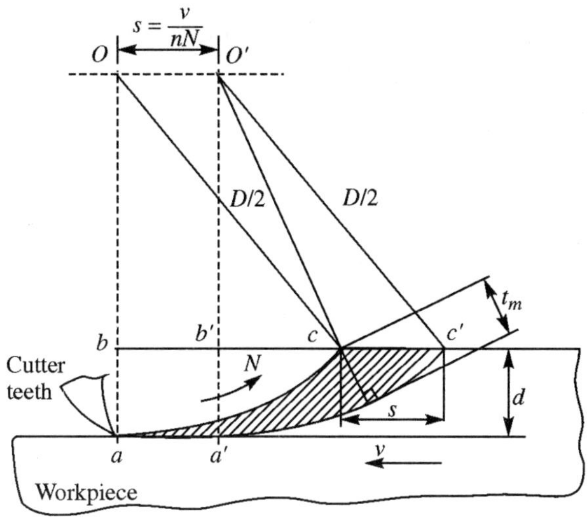

Fig. 10.11 Undeformed chip-thickness in plain milling

The volume of an undeformed chip can also be obtained by dividing the material removed from the workpiece in time τ by the number of chips produced. Thus,

$$(\text{Vol})_{\text{chip}} = \frac{vbd\tau}{nN\tau} = \frac{vbd}{nN} \qquad ...(10.29)$$

where v is the table speed and N is the rpm of the cutter having n number of teeth.

Equating (10.28) and (10.29)., the mean undeformed chip-thickness can be evaluated as

$$\bar{t} = \frac{v}{nN}\sqrt{\left(\frac{d}{D}\right)} \qquad ...(10.30)$$

Approximating the chip section $acc'a'$ by a triangle, the maximum undeformed chip-thickness $t_m \approx 2\bar{t}$.

For applying the orthogonal cutting theory, the rake angle is the same as the radial rake angle (Fig. 10.12) when the helix angle is zero. The chip-thickness, although varies from zero to t_m, may be approximated to be of uniform value equal to $\bar{t}$. This, however, will give only a crude estimate of the forces when used in orthogonal cutting theory of Chapter 4.

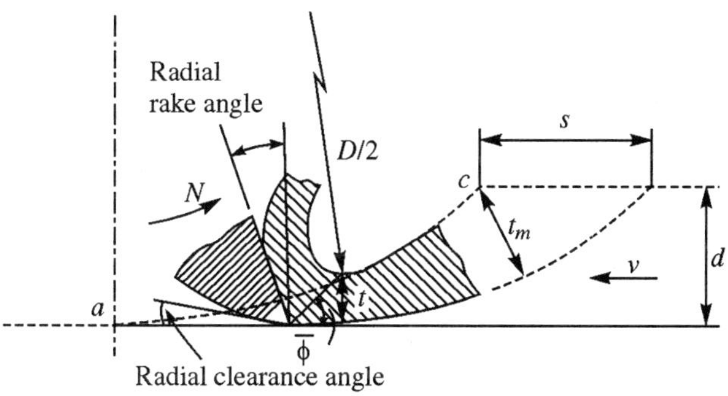

Fig. 10.12 Chip formation in plain milling

In symmetrical face milling, as the tool passes over the workpiece the chip thickness increases to a maximum value and then decreases. Referring to Fig. 10.13, the chip length l can be evaluated as

$$l \approx \text{arc } AB = D\chi \qquad \text{...(10.31)}$$

and

$$\cos\chi = \left[1-\left(\frac{B}{D}\right)^2\right]^{1/2} \qquad \text{...(10.32)}$$

where B is the width of workpiece and D is the cutter diameter. Expanding both sides of equation (10.32) and neglecting higher order terms, we get

$$1-\frac{\chi^2}{2!}+.... = 1-\frac{1}{2}\left(\frac{B}{D}\right)^2-\frac{1}{8}\left(\frac{B}{D}\right)^4-...$$

or

$$\chi = \frac{B}{D}\left[1+\left(\frac{B}{2D}\right)^2\right]^{1/2} \qquad \text{...(10.33)}$$

and

$$l = B\left[1+\left(\frac{B}{2D}\right)^2\right]^{1/2} \qquad \text{...(10.34)}$$

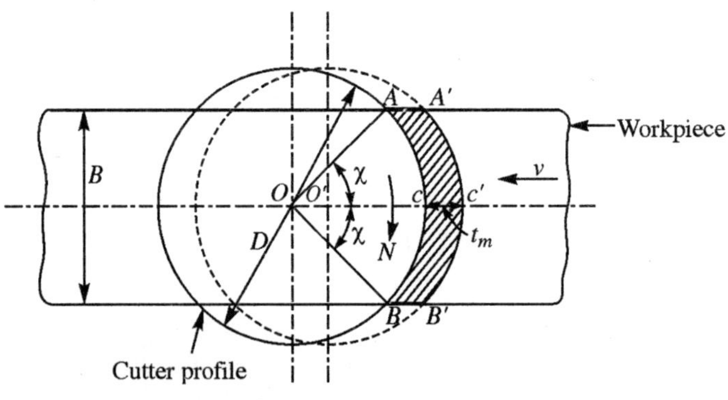

Fig. 10.13 Undeformed chip-thickness in face milling

If $\bar{t}$ is the average thickness of the chip and d is the depth of cut, then the volume of an individual chip can be evaluated as

$$(\text{Vol})_{\text{chip}} = \bar{t}\,d\,B\left[1+\left(\frac{B}{2D}\right)^2\right]^{1/2} \qquad \text{...(10.35)}$$

Dividing the material removed from the workpiece in time τ by the number of chips produced, we get

$$(\text{Vol})_{\text{chip}} = \frac{v\,d\,B\,\tau}{n\,N\,\tau}$$

$$= \frac{v\,d\,B}{n\,N} \qquad \text{...(10.36)}$$

where v is the table speed and N is the rpm of cutter having n number of teeth. Equating (10.35) and (10.36), the average undeformed chip-thickness can be evaluated as

$$\bar{t} = \frac{v}{nN}\left[1+\left(\frac{B}{2D}\right)\right]^{-1/2} \qquad ...(10.37)$$

Expanding $\left[1+\left(\frac{B}{2D}\right)^2\right]^{-1/2}$ and neglecting higher-order terms, we can write

$$\bar{t} = \frac{v}{nN}\left[1-\frac{1}{2}\left(\frac{B}{2D}\right)^2\right] \qquad ...(10.38)$$

It is clear from Figs. 10.11 and 10.13 that in milling operations the undeformed chip size or the size of cut taken by an individual tooth is of prime importance and not the depth of cut d or the table speed v. The size of cut, however, depends on the various cutting parameters.

The tool angles in face milling cutters are similar to those on turning tools except that the nomenclature used are different. These angles are shown in Fig. 10.14. When the helix or axial rake angle is zero, orthogonal cutting conditions may be approximated by taking projections of the mean chip thickness $\bar{t}$ and the radial rake angle on a plane perpendicular to the cutting edge. When the side cutting-edge angle is also zero, α is equal to the radial rake angle.

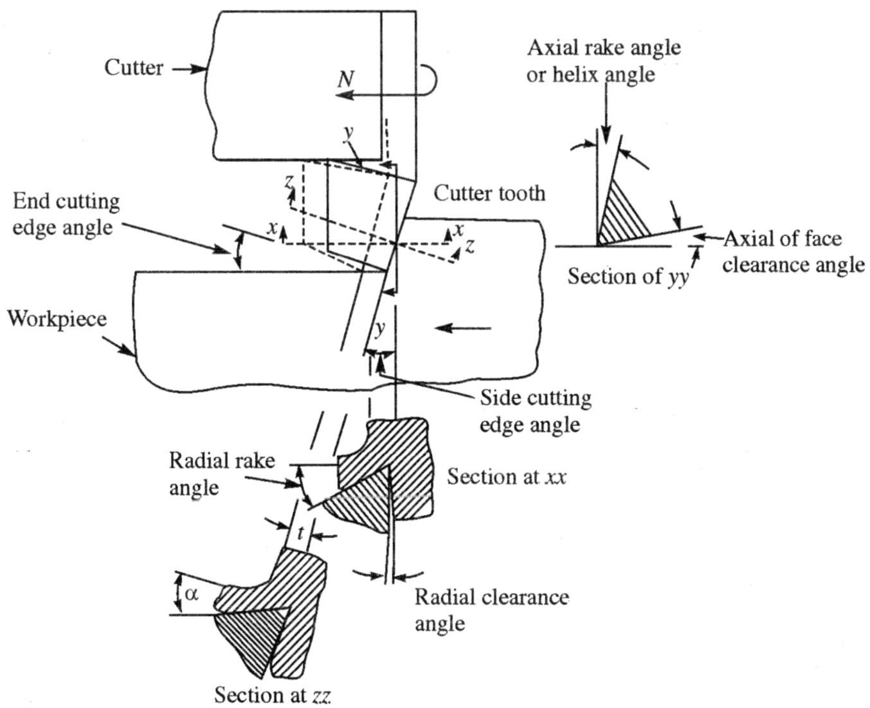

Fig. 10.14 Chip formation in face milling

10.4.2 Forces

Figure 10.15 shows the forces acting during plain milling. The resultant force on the milling cutter can be resolved into three components, F'_p, F'_q and F'_r, acting along the tangential, radial and axial directions as shown in Fig. 10.15(a). Alternatively, the components can be taken as F_p, F_q and F_r along the longitudinal, vertical and transverse directions as shown in Fig. 10.15(b). These directions are useful when a workpiece dynamometer is used for measurement of forces. The axial or transverse component is usually small in comparison to the F_p or F'_p component. When a straight cutter (helix angle = 0) is used, the axial or transverse component is zero and a two-dimensional force system is obtained (Fig. 10.15 c and d).

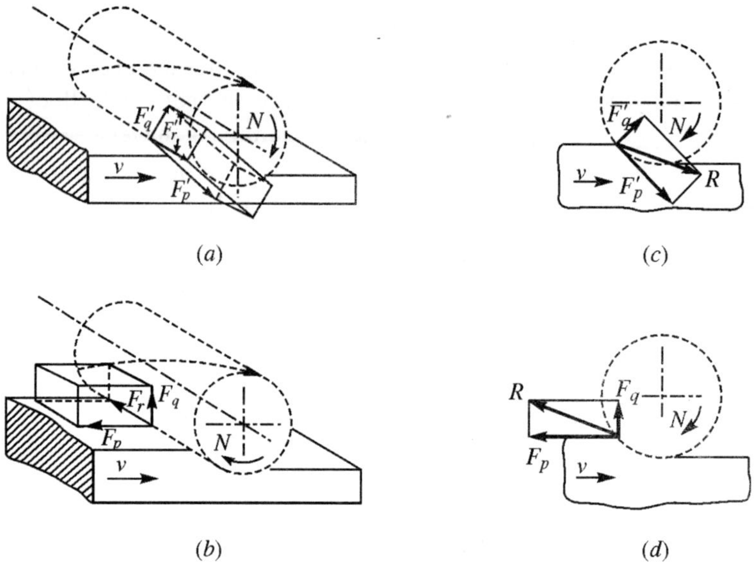

Fig. 10.15 Force components during plain milling

Because of the periodic nature of chip formation in milling, the forces also vary periodically leading to variation in cutting speed and vibrations. This variation in forces have a damaging effect on tool life as well as surface finish.

Figure 10.16 shows how the force varies with the rotation of cutter. Three cases are shown. In the first case, the spacing between teeth on the cutter is greater than the chip length, while in the second and the third cases, the spacing is equal to and less than the chip length, respectively. The resultant pulses in these figures have been obtained by appropriately superimposing the individual pulses. It is clear that the force fluctuations can be reduced by increasing the number of teeth on the cutter but there is a limit to this since sufficient spacing must be available between two teeth to accommodate the chip. The reduction in force fluctuations can also be achieved by using helical cutters (Fig. 10.17) where several teeth remain in contact with the workpiece during cutting. Further, the contact between the workpiece and the cutter in this case starts and ends gradually leading to reduction in shock loads. When the helix angle is zero, the cutting force increases to a maximum value and rapidly drops to zero as shown.

In face milling operations, the force components F_p, F_q and F_r are usually taken along the longitudinal, vertical and transverse directions, respectively.

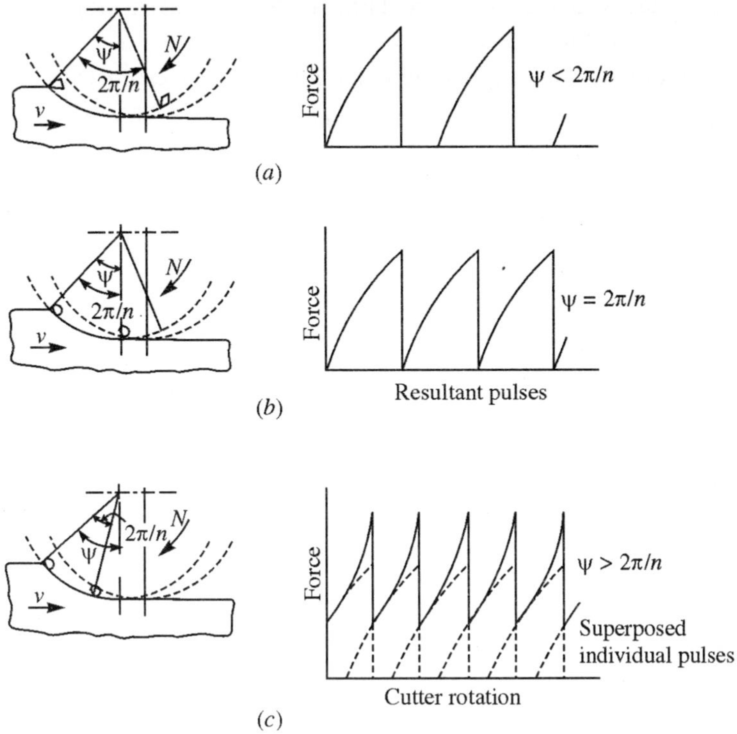

Fig. 10.16 Force fluctuations in plain milling when helix angle is zero

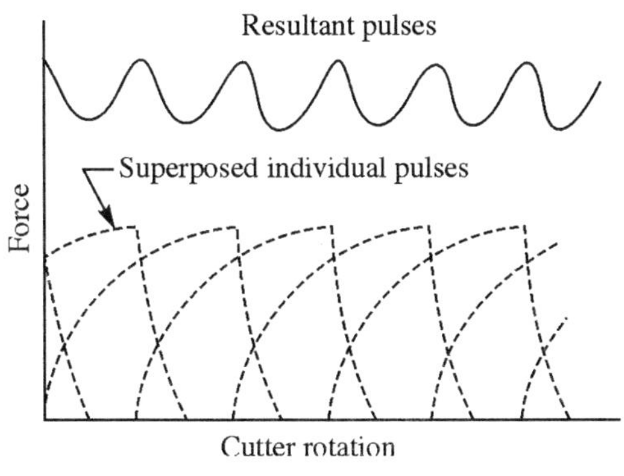

Fig. 10.17 Effect of helix angle on force fluctuations in plain milling

These force components are shown in Fig. 10.18 and can be easily measured with the help of a three-component workpiece dynamometer such as the one discussed in Chapter 6. The transverse component F_r is usually small and depends upon the width of the workpiece. When the workpiece width is small compared to the cutter diameter, this component is negligible. Force fluctuations occur during face milling operations also because of intermittent contact with the workpiece and the variable chip-thickness. Here also the force pulse increases rapidly to a maximum value and drops to zero on disengagement with the workpiece. The force fluctuations can, however, be reduced by increasing the number of teeth in contact with the workpiece during cutting. Initial shock is reduced by providing the cutting-edge angle as on turning tools.

When the helix or axial rake angle is zero, the forces can be estimated using orthogonal cutting model by assuming a constant undeformed chip-thickness equal to $\bar{t}$. This of course requires projections on appropriate planes as shown in Fig. 10.14.

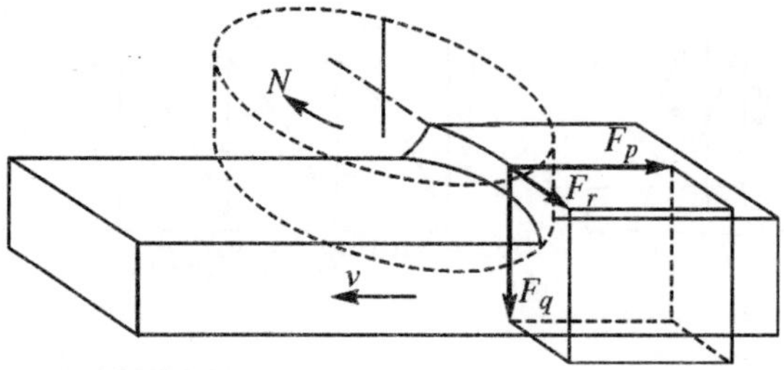

Fig. 10.18 Force components during face milling

10.4.3 Surface Finish

Ideal surface roughness in plain milling can be easily evaluated from Fig. 10.19. It is generally found that superimposed waviness occurs for each revolution of the cutter. This waviness may be due to eccentricity in the spindle or variation in tooth heights. For ideal surface roughnes, the surface generated by an individual tooth may be approximated by an arc of a circle of diameter equal to that of the cutter. The surface generated will also have the feed marks by individual tooth as shown.

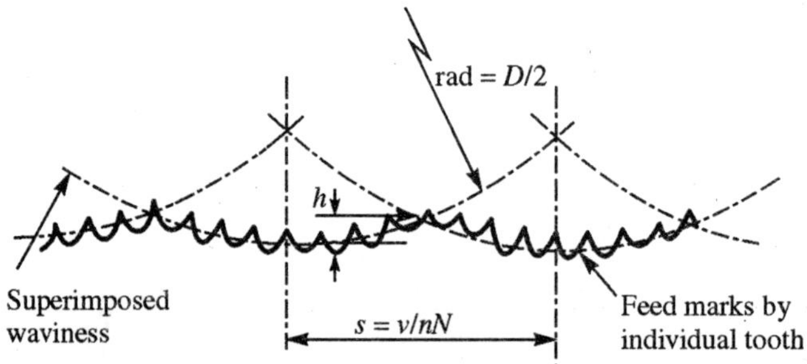

Fig. 10.19 Surface roughness in plain milling

From Fig. 10.19, we can write

$$\left[\left(\frac{D}{2}\right)^2 - \left(\frac{D}{2} - h\right)^2\right] = \left(\frac{s}{2}\right)^2 \qquad \text{...(10.39)}$$

where s is the feed per tooth. On simplification, this equation gives

$$h = \frac{v^2}{4Dn^2N^2} \qquad \text{...(10.40)}$$

Ideal surface roughness in face milling can be obtained from the feed marks left behind by the cutter teeth. The general approach for evaluating h is similar to that followed for turning since the geometry of face milling cutter tooth is similar to that of a turning tool. Approximating the path followed by an individual tooth to be an arc of a circle having radius equal to that of the cutter, the feed marks can be drawn as shown in Fig. 10.20. The peak-to-valley height can now be estimated from

$$h = \frac{s}{\tan\gamma_s + \cot\gamma_e} \qquad ...(10.41)$$

where s is the feed per tooth, γ_e is the end cutting-edge angle and γ_s is the side cutting-edge angle. These angles are shown in Fig. 10.14.

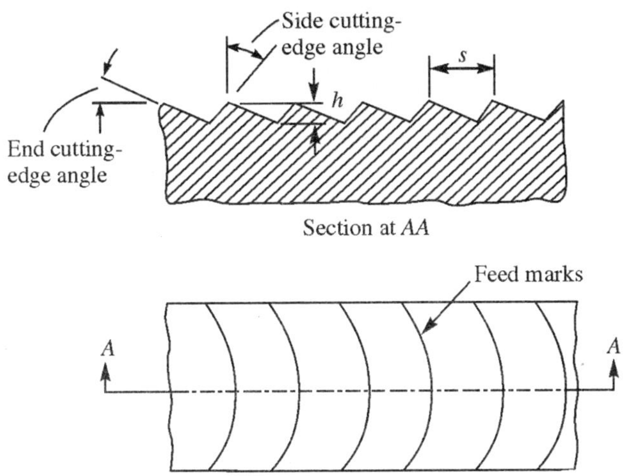

Fig. 10.20 Surface roughness in face milling

10.5 DRILLING

The geometry of a twist drill was discussed in Chapter 3. Here cutting occurs on the straight cutting edges (cutting lips) and the chisel edge. The chip formation, however, occurs mainly at the cutting edges and the action at the chisel edge is one of pushings into the material like a wedge. The chips generated at the cutting edges move up the axis of the drill along the flute. When the drill is fed downwards with feed f, each of the two lips remove a layer equal to $f/2$ as shown in Fig. 10.21. The normal rake angle α_n can be obtained by taking a section along AA, perpendicular to the cutting edge. Angle α_n varies with distance r_a and should, therefore, be defined with reference to the radial location on the section and can be evaluated approximately from

$$\tan\alpha_n \approx \frac{r_a \tan\psi}{\frac{D}{2}\sin\frac{\beta}{2}} \qquad ...(10.42)$$

where β and ψ are the point and helix angles and D is the twist drill diameter. The helix angle varies along the length of the cutting edge but its value ψ is usually defined at the periphery. At the middle of the cutting edge, *i.e.*, at $r_a = D/4$, the normal rake angle becomes

$$\alpha_n \approx \tan^{-1}\left[\frac{\tan\psi}{2\sin\beta/2}\right] \qquad ...(10.43)$$

The cutting speed and other parameters also vary considerably along the cutting edge making the chip formation process very complex to analyse. Drilling is essentially an oblique cutting operation

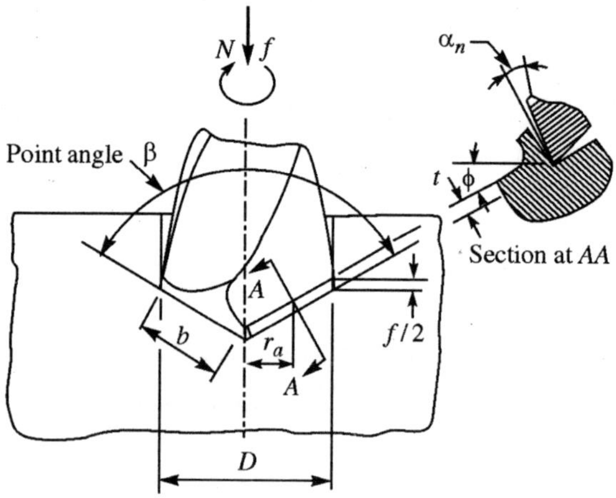

Fig. 10.21 Chip formation in drilling

10.5.1 Undeformed Chip-thickness

Referring to Fig. 10.21, the undeformed chip-thickness t and the chip width b can be evaluated as

$$t = \frac{f}{2}\sin\beta/2 \qquad \text{...(10.44)}$$

and

$$b = \frac{D}{2\sin\beta/2} \qquad \text{...(10.45)}$$

10.5.2 Forces and Power

The drilling force can be resolved into the components F_p, the tangential component, and F_q', the force component normal to the cutting edge. F_q' can be further resolved into two components, the radial component F_r and the thrust component F_q acting axially. These forces are shown in Fig. 10.22.

The total force F_p in the tangential direction during drilling not only consists of the tangential components at the cutting edges but the forces due to flank friction and chisel edge also contribute to it. Thus, for a twist drill with two lips

$$F_p = 2[F_p + F_{p_1} + F_{p_2}] \qquad (10.46)$$

where F_{p_1} and F_{p_2} are the tangential components of the forces due to friction and extrustion at the chisel edge. Similarly, the total thrust force F_Q during drilling with a twist drill having two lips is given by

$$F_Q = 2[F_q + F_{q_1} + F_{q_2}] \qquad (10.47)$$

where F_{q_1} and F_{q_2} are the thrust forces due to friction and extrusion.

Experiments show that F_{p_1} and F_{p_2} are usually much smaller than F_p with more than 85% of the torque coming from cutting action at the lips. Also the contribution of force due to friction F_{q_1} on total thrust is negligibly small, usually less than 3%.

Empirical relations based on experimental results are of the form

$$F_p = C_1 f^{x_1} D^{y_1} \quad \text{...(10.48)}$$

and

$$F_q = C_2 f^{x_2} D^{y_2} \quad \text{...(10.49)}$$

where exponents x_1, y_1, x_2 and y_2 and constants C_1 and C_2 can be determined experimentally for a particular tool-work combination. For mild steel, the value of exponents x_1, y_1, x_2 and y_2 are generally around 0.8, 2.0, 0.8 and 1.0, respectively.

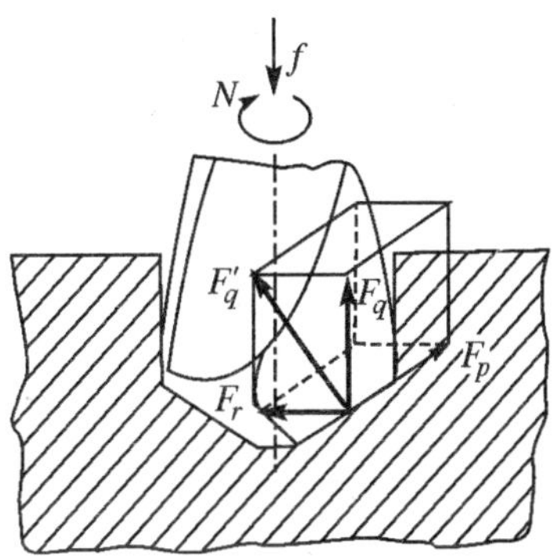

Fig. 10.22 Force components during drilling

10.5.3 Surface Finish

Drilling operations are usually considered as roughing operations and not much concern is shown on the quality of surface finish produced. For better finish and accuracy, the drilling operation is usually followed by reaming.

CHAPTER 11

Grinding of Metals

11.1 INTRODUCTION

In the grinding process, material removal is achieved by means of a rotating abrasive wheel in which a large number of abrasive grains are held together with the help of a bonding agent. Grinding and other machining processes are basically the same to the extent that in both cases metal is removed by the process of plastic flow and the deformed metal leaves the workpiece in the form of chips. The mechanics of the grinding process are, however, difficult to observe and evaluate due to random geometry and distribution of the cutting points, high cutting speed and small undeformed chip-thickness. It gets further complicated because the geometry of cutting points or abrasive grains not only varies from one grain to another but also changes continuously and unpredictably due to wear during grinding. Thus, parameters such as tool geometry, chip-thickness, etc. which are so important for understanding the single-point cutting operations cannot be evaluated for individual grains but only an average can be considered.

11.2 THE GRINDING WHEEL

For manufacturing grinding wheels, the abrasive grains are mixed with a bounding agent and pressed or cast into the required shape and baked in kilns or electric furnaces. The performance of a grinding wheel depends upon, (*i*) the type of abrasive grains, (*ii*) the size of abrasive grains, (*iii*) the type and quantity of bonding material, and (*iv*) the spacing between the grains.

11.2.1 Abrasives

Abrasives which are widely used in grinding wheels are aluminium oxide (Al_2O_3) and silicon carbide (SiC). Aluminium oxide grains are suitable for grinding of steels, ferrous alloys and other high-strength materials, while silicon carbide grains are used for grinding of hard and brittle materials such as chilled cast irons, ceramics, glass as well as other low-strength materials such as aluminium, brass and copper. Aluminium oxide grains when used on hard and brittle materials tend to glaze rather than cut. Aluminium oxide grains being less brittle are preferred for grinding tougher materials. Silicon carbide grains when used on tougher materials fracture readily resulting in excessive wheel wear. Diamond grains are used for grinding extremely hard materials such as tungsten carbide and ceramics.

A variety of aluminium oxide grains are used in grinding wheels, the pure form of fused aluminium oxide being the most common one. White aluminium oxide is a more refined form and has high friability. Sometimes oxides such as titanium oxide, chromium oxide or zirconium oxide are added to vary its characteristics. Similarly, a variety of silicon carbide grains have been produced depending upon the crushing, shaping and chemical treatment steps taken in reducing the crude silicon carbide to abrasive grain stage. The black and the green silicon carbide varieties are the most common ones. Boron carbide (B_4C) and cubic boron nitride (CBN) are the other two abrasive materials which are used in grinding wheels. Both of these materials are extremely hard with hardness values approaching to that of diamond. Cubic boron nitride wheels are particularly used for high-speed grinding and for grinding of extra hard and difficult-to-grind materials primarily because it has high thermal stability and is inert to chemical reactions.

11.2.2 Grain Size

The size of an abrasive grain is expressed in terms of a sieve number (s_n) which corresponds to the number of openings per linear inch. The mean diameter of an abrasive grain (g) is related to the sieve or screen number approximately by $g = \dfrac{0.6}{s_n}$ (inch).

Thus, a number 30 grain size has a nominal diameter of about 500 microns, while a 46 grain size has about 325 microns.

11.2.3 Bond Materials and Structure

A variety of bond materials have been used for bonding the grains. They provide the required strength and hardness to the grinding wheels. The bonding should be strong enough to withstand the stresses at the grinding temperatures. It should provide adequate support to the abrasive grains during cutting but should be friable enough to dislodge the worn or dull grains. The bond materials should also be resistant to chemical attack by the coolant. Considerations of these have resulted in the use of widely different bond materials, each satisfying a particular requirement. The common ones are vitrified, resinoid, rubber, shellac and silicate bonds, vitrified being the most commonly used in general engineering industries.

The volumetric compositon of a grinding wheel can be expressed as

$$V_a + V_b + V_p = 100\%$$

where V_a, V_b and V_p are the volume percentage of abrasive grains, bond material and pores (voids), respectively. In a typical vitrified wheel, $V_a \approx 50\%$, $V_b \approx 10\%$ and $V_p \approx 40\%$. These three constituents can be altered to control the structure and grade (hardness) of a grinding wheel.

The structure of a grinding wheel is a measure of the dispersion of abrasive grains in the wheel and characterises the mean void size and grain distribution. The structure is designated by a number, 1 representing a very dense structure and 16 a very open one. Thus, in a dense or closely-packed structure, V_a is high and it is low in an open or loosely-packed structure.

11.2.4 Grade

The grade or hardness of a grinding wheel is a measure of the ability of the bond material to retain the grains. The property is characterised by the mean force required to dislodge a grain from the surface of a wheel. The grade of a wheel is designated by a letter of alphabet, *A* at the soft end of the scale and *Z* at the hard end. A soft wheel readily loses its grains, while grains on a hard wheel are difficult to dislodge. For a given structure, a harder wheel, therefore, has a higher percentage of bond material and a smaller percentage of voids.

11.2.5 Wheel Specification

The complete specification for conventional grinding wheels is given in an alphanumeric system which is a collection of letters and numbers. This is explained in Fig. 11.1. This marking system for grinding wheels was recommended by ISO and has been adopted by almost all countries. Indian specification also conforms to this system. In this system grain type is denoted by a letter to which a prefix may be added to denote special characteristics. For example, A stands for regular aluminium oxide but AA denotes white aluminium oxide. Some manufacturers prefer numbers as prefix instead of letters. Similarly, a suffix invariably in the form of a number is attached to bond type to indicate special characteristics.

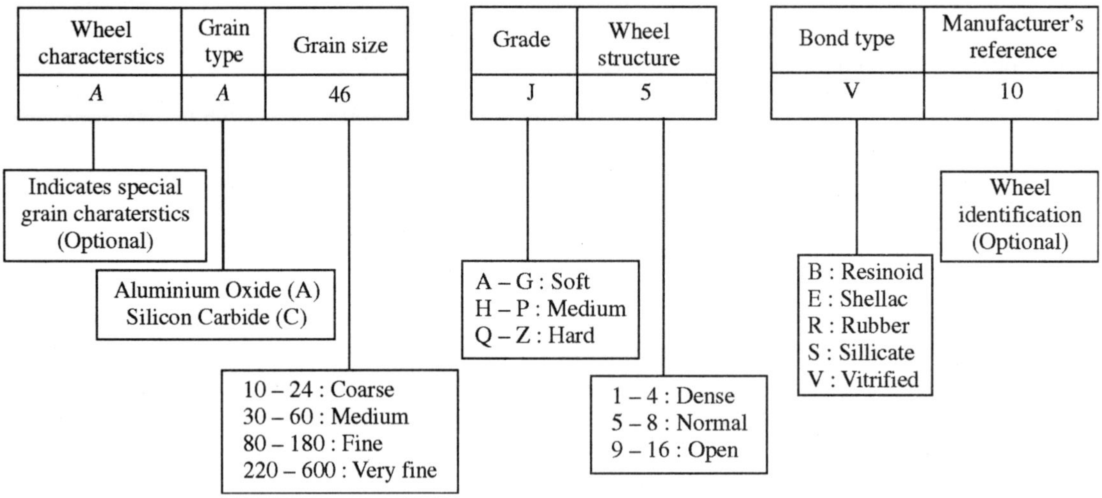

Fig. 11.1 Grinding wheel specification

A wheel marked AA46-J5-V8 indicates a grinding wheel with white aluminum oxide grains (AA) of medium grain size (46), medium grade (J) and normal structure (5) bonded together with a vitrified bond (V) and suitable for precision grinding (8).

11.3 MECHANICS OF GRINDING PROCESS

In grinding, metal is removed by abrasive particles on the wheel surface in the form of chips. The thickness of these chips play a very important role. The area of contact between the chip and the grinding depends on chip thickness and determines the force on an individual grain and hence the bond strength and wear characteristics. In addition, the energy required to remove a unit volume of material is influenced by chip size and this in turn dictates the grinding temperature and hence has a strong influence on the rate of wear and workpiece surface damage. The surface roughness is also a composite effect of individual scratches.

In view of the important role the chip thickness plays in grinding, its value should be known for various grinding conditions. Let us consider the cases of horizontal surface grinding, external cylindrical grinding and internal grinding. The kinematic arrangements for these grinding processes are shown in Fig. 11.2. In each case, the wheel speed is designated by V. work speed by v and infeed by d.

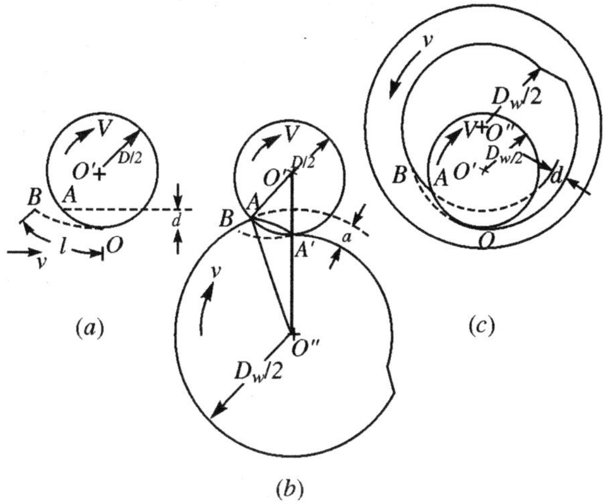

Fig. 11.2 Chip formation during (a) Horizontal surface grinding (b) External cylindrical grinding and (c) Internal grinding

11.3.1 Chip Length

Referring to Fig. 11.2, the undeformed chip length (l) for various cases can be estimated as

$$l = OB \approx OA + AB \qquad ...(11.1)$$

If τ is the time taken for the wheel to travel the distance OA, then

$$OA = V\tau \qquad ...(11.2)$$

and

$$AB = v\tau \qquad ...(11.3)$$

Eliminating τ, we get

$$AB = \frac{v}{V}.OA \qquad ...(11.4)$$

and

$$l = \left(1+\frac{v}{V}\right).\ OA \qquad ...(11.5)$$

Approximation of arc OA by the chordal distance gives

$$OA^2 = \left[\left(\frac{D}{2}\right)^2 - \left(\frac{D}{2}-d\right)^2\right] + d^2$$

or

$$OA = \sqrt{(dD)} \qquad ...(11.6)$$

for horizontal surface grinding. The chip length l_s for this case becomes

$$l_s = \left(1+\frac{v}{V}\right)\sqrt{(dD)} \qquad ...(11.7)$$

For external cylindrical grinding, the chordal length OA can be evaluated using triangles $O'AA'$ and $O''AA''$ (Fig. 11.2b). OAA', we get

$$OA^2 = \left[\left(\frac{D}{2}\right)^2 - \left(\frac{D}{2} - OA'\right)^2\right] + OA'^2 = D.OA' \qquad ...(11.8)$$

Also,

$$OA^2 = \left(\frac{D_w}{2}\right)^2 - \left(\frac{D_w}{2} - d + OA'\right)^2 + OA'^2$$

$$= dD_w - 2\left(\frac{D_w}{2} - d\right)OA' \qquad ...(11.9)$$

In these equations, D as before is the wheel diameter and D_w is the work diameter. Substituting for OA' from equation (11.8) and simplifying, we get

$$OA \approx \sqrt{\left[\frac{dD}{(1 + D/D_w)}\right]} \qquad ...(11.10)$$

Hence, chip length l_e for external cylindrical grinding is obtained as

$$l_e = \left(1 + \frac{v}{V}\right)\sqrt{\left[\frac{dD}{(1 + D/D_w)}\right]} \qquad ...(11.11)$$

The chip length l_i for internal cylindrical grinding can be similarly obtained as

$$l_i = \left(1 + \frac{v}{V}\right)\sqrt{\left[\frac{dD}{(1 - D/D_w)}\right]} \qquad ...(11.12)$$

When work speed v is much smaller than the wheel speed V, which is usually the case in peripheral grinding operations, equations (11.7), (11.11) and (11.12) can be written as

$$l_s = \sqrt{(dD)}$$

$$l_e = \sqrt{\left[\frac{dD}{(1 + D/D_w)}\right]}$$

and

$$l_i = \sqrt{\left[\frac{dD}{(1 - D/D_w)}\right]} \qquad ...(11.13)$$

11.3.2 Chip thickness

The chip thickness as mentioned earlier, is an important variable affecting the grinding characteristics. It significantly influences the grinding energy, wheel wear, surface finish, etc. Hence, it is important to predict the magnitude of this undeformed chip-thickness. This is usually obtained by equating the mean volume of a single chip estimated in two different ways. It is assumed that all chips are of the same average size and that the chips have a constant width b. A typical case of horizontal surface grinding is shown in Fig. 11.3.

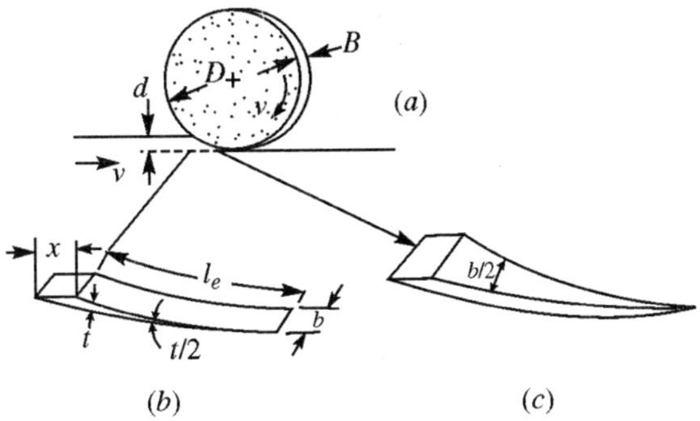

Fig. 11.3 Chips produced in horizontal surface grinding (a) Schematic view of the process (b) Idealized chip and (c) Actual chip

Here the chip length l_s is very much larger than the maximum undeformed chip-thickness t. The volume of an individual chip is

$$(\text{Vol})_{\text{chip}} = \frac{1}{2} l_s t b = \frac{l_s t^2 r}{4} \qquad \text{...(11.14)}$$

where

$$r = \frac{b}{\frac{1}{2}t} \qquad \text{...(11.15)}$$

The chip volume can also be obtained by dividing the rate of metal removal by the number of cuts made per unit time. This assumes that during wheel-work interaction, each grain produces a chip. Thus,

$$(\text{Vol})_{\text{chip}} = \frac{vdB}{VCB} = \frac{vd}{VC} \qquad \text{...(11.16)}$$

where C is the number of active grains per unit area and B is the width of grinding wheel in contact with the workpiece. Equating the chip volumes, we get

$$t^2 = \frac{4v}{VCr} \cdot \frac{d}{l_s} \qquad \text{...(11.17)}$$

Substituting for l_s from equation (11.13), we obtain t for surface grinding as

$$t = \left[\frac{4v}{VCr} \sqrt{\left(\frac{d}{D} \right)} \right]^{1/2} \qquad \text{...(11.18)}$$

The value of t for all types of peripheral grinding operations can be obtained in a similar manner. Replacing l_s by l_e and l_i, the value of t for external and internal cylindrical grinding can be obtained and the expression for t can be written in a general form as

$$t = \left[\frac{4v}{VCr} \sqrt{\left\{ \frac{d(1 + D/D_w)}{D} \right\}} \right]^{1/2} \qquad \text{...(11.19)}$$

where D_w is infinite for surface grinding, positive for external cylindrical grinding and negative for internal grinding.

11.3.3 Cutting Points

The simplest way of determining *C*, the number of cutting points on the wheel surface, is to roll a dressed wheel over a soot-coated plate and count the number of contacts indicated on an enlargement made using the plate as a negative. This is, however, a crude method and invariably gives a rather high value for *C*. Another method for evaluating *C* is to count the number of chips produced in a given time. This method assumes that each contact point on the wheel removes a chip at each engagement with the workpiece. A better method and a more successful one for determining *C* is to count the number of wear flats produced on the wheel surface. For this purpose, the wheel surface is lighted at a glancing angle and observed under a microscope. Experimental observations show that the dynamic or active grain density (number of cutting points per unit area) initially increases steeply as the chip thickness is increased and then tends to remain more or less constant as shown in Fig. 11.4. Thus, at higher values of chip thickness, *i.e.*, in the practical grinding range, the active grain density is likely to be independent of the grinding conditions.

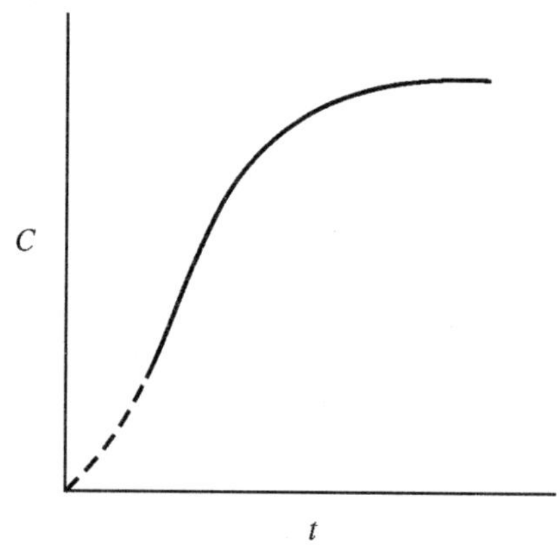

Fig. 11.4 Variation of active grain density *C* with *t*

11.4 GRINDING FORCES AND SPECIFIC ENERGY

The forces acting at the wheel-work interface can be easily measured with the help of a two or three component workpiece dynamometer. Figure 11.5 shows the three force components F_p, F_q and F_r during horizontal surface grinding. When the workpiece width B' is less than the width of the grinding wheel *B*, no cross-feed is needed and F_r component becomes zero.

The specific energy *u* in grinding (energy required to remove a unit volume of material) can be evaluated from force measurements and volume of material removed. From measured values of grinding forces F_p, the specific energy values can be calculated using the equation

$$u = \frac{F_p V}{vBd} \quad ...(11.20)$$

This specific energy in grinding has been found to be an order of magnitude higher than that for single-point cutting such as turning. It has been shown experimentally that it increases rapidly when the chip-thickness *t* is decreased as shown in Fig. 11.6. This increase in specific energy with decreasing chip thickness has been attributed to the 'size effect' analogous to the increase in breaking strength of very fine wires with decreasing diameter. At very small chip-thickness involved in grinding, the plastic deformation occurs over an area having very few or no imperfections and the material tends to behave like an ideal material with no inhomogeneities. Under these conditions the shear stress required to cause plastic deformation is very much higher and, therefore, the specific energy is high.

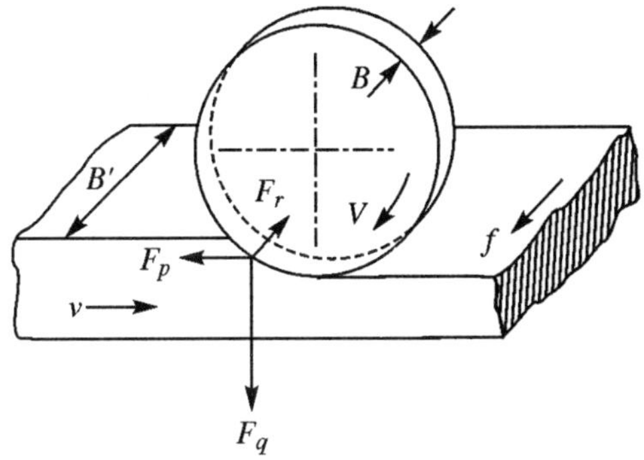

Fig. 11.5 Force components during horizontal surface grinding

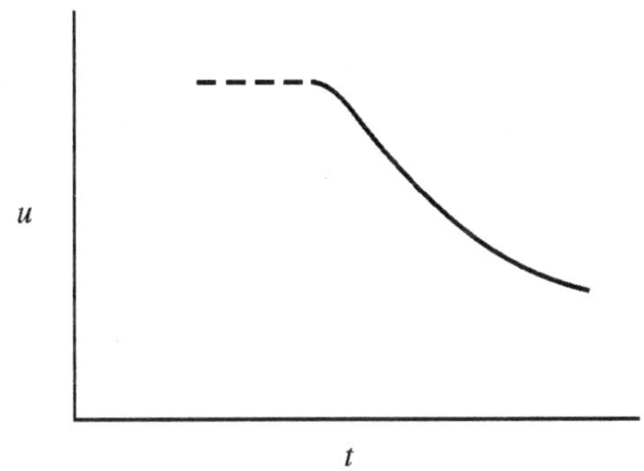

Fig. 11.6 Variation of specific energy *u* with *t*

While size effect is an important factor contributing to high specific energy in grinding, evidence of excessive rubbing and ploughing during grinding have also been presented. During controlled force plunge grinding it has been found that under light loads, there is no material removal and all the energy is consumed in rubbing. When the load is further increased, ploughing occurs where the material is simply pushed to the sides of the grooves generated by the abrasive grains on the wheel. Material removal or cutting occurs only when the load is increased beyond a certain value. These three regions are shown in Fig. 11.7. Similar results have been obtained during cutting experiments with single abrasive grains. Thus, both size effect as well as rubbing and ploughing are significant factors contributing to high specific energy in grinding.

When specific energy values obtained in grinding are plotted against t on a log-log graph, the variation is seen to be linear as shown in Fig. 11.8(a). This gives a relationship of the form

$$u\, t^m = \text{Constant}$$

or

$$u \propto \frac{1}{t^m} \qquad \text{...(11.21)}$$

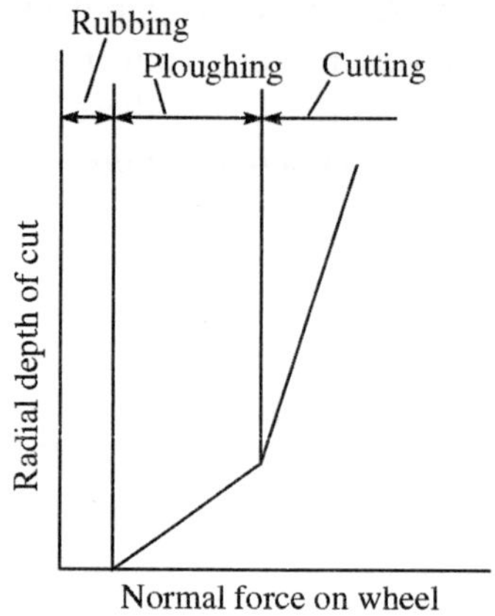

Fig. 11.7 Effect of normal force on wheel on radial depth of cut indicating rubbing, ploughing and cutting regions

In the fine grinding range, the value of exponent m is usually around 0.9 and can be taken to be approximately 1. When specific energy values for fine grinding, rough grinding and single-point cutting are plotted together, a continuous curve such as one shown in Fig. 11.8(b) is obtained. The curve indicates that the exponent m in the rough grinding range is about 0.6 while in the single-point cutting range it is only around 0.2.

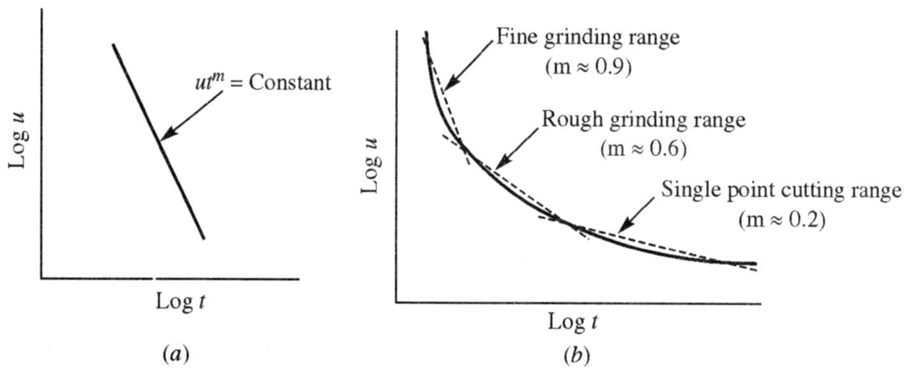

Fig. 11.8 Variation of specific energy u with chip thickness t

The hypothesis that grinding and single-point cutting are analogous has led to the determination of rake angle of a typical abrasive grain. A zero representative rake angle based on the probability of having as many grits with positive as with negative rake angle was assumed earlier. Under plunge-cut grinding conditions the ratio of grinding forces F_p/F_q is generally around 0.5 and is relatively insensitive to changes in grinding conditions. In orthogonal cutting, this force ratio strongly depends on the rake angle of the tool. The ratio decreases from 1.1 to 0.48 as the rake angle is reduced from +10° to –45° (Fig. 11.9). If the analogy with single-point cutting is correct, then the typical abrasive grain should have a large negative rake angle. Tests conducted with single abrasive grain, however, suggest that an average grain in fine grinding may be approximated by a sphere, whereas in rough grinding, the shape may be better approximated by a blunt cone. The manner in which tools having such idealised shapes remove a chip is not clear. Experimental results with spherical and blunt conicial tools show that the force ratio varies significantly when the depth of cut is increased (Fig. 11.10) and that at small depths of cut practically no material removal takes place and ploughing dominates.

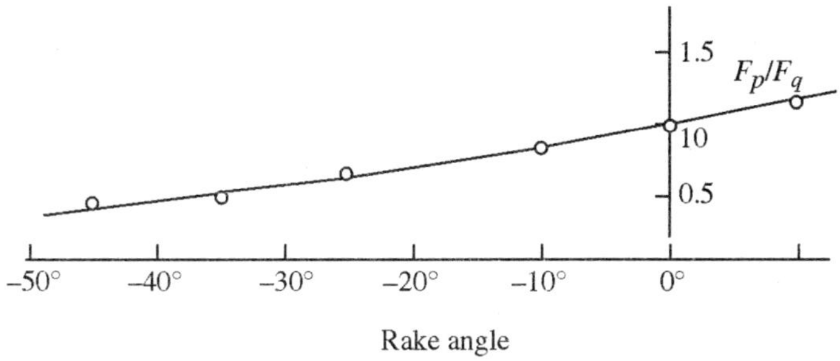

Fig. 11.9 Variation in force ration F_p/F_q with rake angle (After Komanduri, 1971)

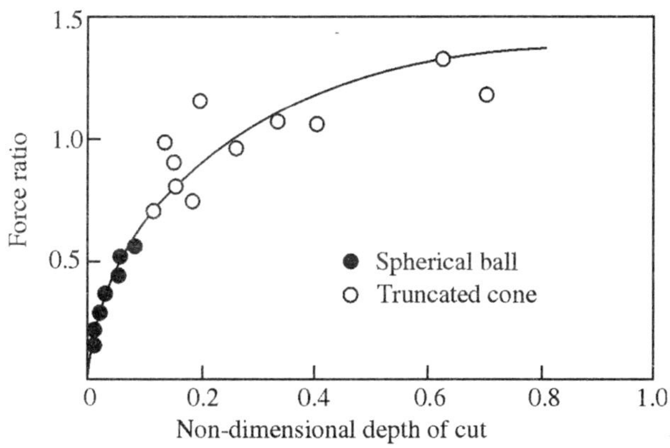

Fig. 11.10 Variation in force ratio with depth of cut during scraping with spherical balls and truncated cones (Data from scrutton *et al.*, 1969 and Lal and Shaw, 1974)

11.5 WHEEL WEAR AND GRINDING PERFORMANCE

Wear pattern of grinding wheel is very similar to that of single-point cutting tools (Fig. 8.2). For fine grinding, the stages of wheel wear (W_w) with respect to material removed (W_m) is shown in (Fig. 11.11). It shows three distinct phases. The first phase indicates rapid wear of the grinding wheel due to wearing of high spots on the wheel surface (profusion of sharp grains). The behaviour of wheel during this phase is strongly dependent on the dressing technique. This phase, however, lasts for a very short time and the wheel quickly attains a stable grinding condition when phase two is reached. Phase two pertains to the linear portion of the curve where rate of wheel wear remains constant. The limit of this phase indicates the useful life of a wheel after which dressing is necessary. In the third phase, the wheel becomes dull due to loading and develops a glazed appearance resulting in reduction of material removal rate. This phase is signified by increase in grinding forces, excessive vibrations and appearance of grinding burns on the ground surface due to high temperature.

The wheel wear pattern in stock removal (rough grinding) processes such as vertical surface grinding, abrasive cut-off operations, etc. would differ from the curve shown in Fig. 11.11 in that the linear portion of the curve (phase two) would extend almost infinitely as shown by the dotted line. This is due to the fact that there is continuous self-sharpening action and hence there is practically no wheel dulling which obviates the need for dressing.

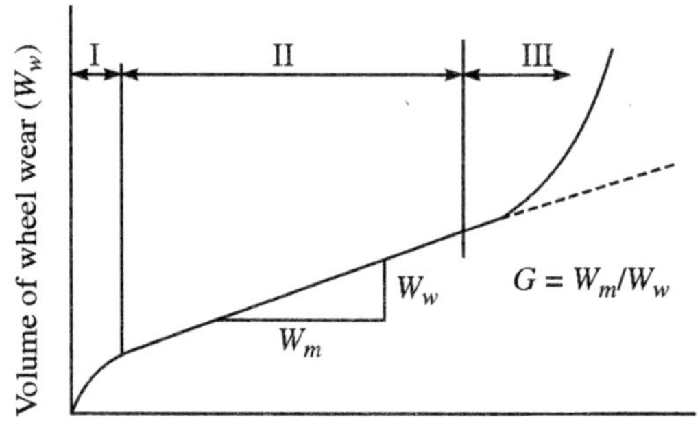

Fig. 11.11 Variation in grinding wheel wear with volume of material removed

In order to understand the wheel wear phenomenon, let us look at the factors responsible for the wear of a grain. Attritious wear occurs at the grain-workpiece contact surface (*a*) as shown in Fig. 11.12 and is gradual wearing away of the sharpness of the cutting points resulting in the formation of wear flats on the grain tips. Attritious wear is due to micro-fracture of cutting points, plastic flow at the grain tips and crumbling of sharp edges. These get aggravated due to high temperatures obtained in grinding. When the stresses on the grain increase to a sufficiently high value, the grain fractures into fragments and fresh cutting edges are exposed (*b*). This is called *fracture wear*. When the grain is stronger than the bond-post, as in the case of softer wheels, the grains get pulled out of the bond-post when sufficient force is applied at the grain tip. Generally the interface between the grain and the bond-post is quite strong, and hence the pulling out of the grain would be due to fracture of the bond-post (*c*) rather than separation at the interface. The wear due to this is generally called the *bond-post fracture wear*. A typical sequence of events, therefore, comprise, of dulling due to attrition and then fracturing away of a small portion of the grain followed by breaking of the bond-post and final release of the remainder of the grain from the wheel. Glazing occurs when worn abrasive grains are not fractured away or dislodged from the wheel. By increasing the force on an individual grain, the worn grit can be fractured or pulled out of the bond post. This will expose new grains and restore the cutting ability of the wheel. Similarly, when the wheel wear is high, the force on an individual grain should be reduced.

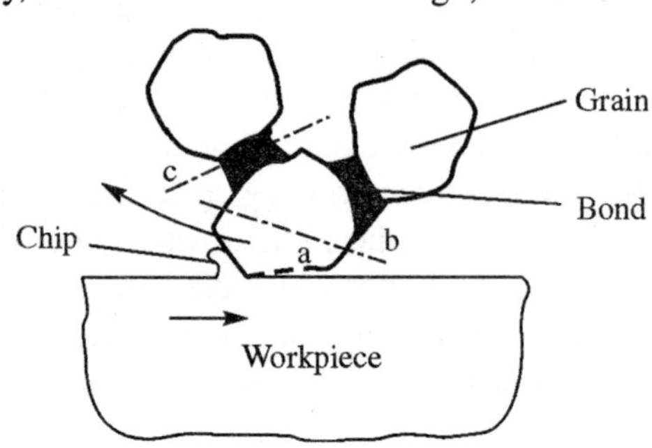

Fig. 11.12 Illustration of three types of wear. a - Attritious wear, b -Grain fracture and c - Bond fracture

Assuming the tangential force F_g on an individual grain to be proportional to be area of cut A_c, we can write that

$$F_g \propto A_c$$
$$\propto b\,.\,t$$
$$\propto t^2$$

or

$$F_g = K_g\left(\frac{v}{VC}\right)\sqrt{\left[\frac{d(1+D/D_w}{D}\right]} \qquad ...(11.22)$$

where K_g is a constant. For surface grinding, we get F_g as

$$F_g = K_g\left(\frac{v}{VC}\right)\sqrt{\left(\frac{d}{D}\right)} \qquad ...(11.23)$$

These equations show how the variables should be changed to alter the forces acting on an average grain. If the wheel wear is excessive, v and d should be decreased and to prevent wheel glazing v and d should be increased. Similarly, in external cylindrical grinding operations, a wheel will appear softer when the work diameter D_w is decreased. The reverse will be the case for internal grinding.

Experiments show that chemical reactions between abrasive grains and workpiece sometimes have a very significant effect on the wear phenomenon. The presence of high temperature at the interface also contributes to it and some researchers have suggested that attritious wear could be regarded as a thermally activated mechanism. Attritious wear, although very important from surface quality considerations, forms a very small fraction of the total wear volume and the loss of wheel during grinding is mainly due to grain fracture and fracture at the bond posts.

The method of evaluating wheel wear has also been a subject of discussion. Earlier, the wheel wear was evaluated from the reduction in wheel diameter. Experiments have shown that the average size of abrasive particles that come out of the wheel are usually larger than the reduction in wheel diameter, the method therefore may give erroneous values for wear volume. Wheel wear values computed from the abrasive particles separated from the debris collected during grinding have been used in subsequent experimental analyses.

Comparison between the size distribution of abrasive particles in the debris and the size distribution of the grains used in manufacturing the wheel gives an indication of the nature of wheel wear. A typical result for fine grinding is shown in Fig. 11.13. Here the wear particles are smaller than the original grain sizes indicating sufficient grain fracture. If we assume that all larger particles are generated by bond fracture only, then the percentage of bond fracture can be obtained as the ratio of weight of larger wear particles to the total wear particles. Analysis indicates that bond post fracture in this case was around 60%.

The performance of a grinding wheel is usually evaluated in terms of the grinding ratio (G) defined as the ratio of volume of material removed to the volume of wheel wear. The representative value is, however, taken during the stable grinding region (phase II) where the volume of wheel wear increases linearly with increase in the volume of material removed. The initial wheel wear, which is of short duration and not linear in nature, is not considered in evaluating the grinding ratio. The grinding ratio can, therefore, be easily obtained from the slope of curve in phase II as shown in Fig. 11.11.

A typical experimental result showing the variation of wheel wear with number of passes during horizontal surface grinding is shown in Fig. 11.14(a). Figure 11.14(b) shows the corresponding variation in grinding forces. In these figures, phase I and phase II are clearly distinguishable and it is obvious that phase III has not been reached. These figures further indicate that the wheel is cutting efficiently during phase II where wheel wear increases linearly and forces remain more or less constant.

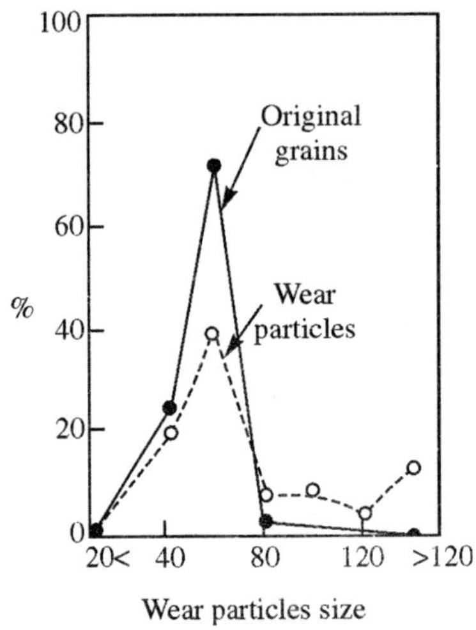

Fig. 11.13 Size distribution of wear particles 46 - *J* wheel, 60 R_B steel (After Pande *et al.*, 1980)

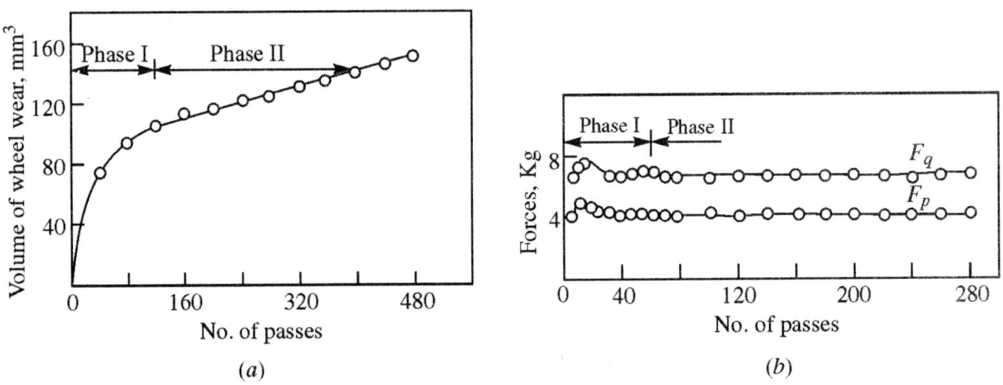

Fig. 11.14 Variation in wheel wear and forces with number of passes (After Pandey *et al.*, 1980)

The variation in grinding ratio with wheel grade for three different work materials is shown in Fig. 11.15(a). In all cases *G* increases initially as wheel hardness is increased but decreases with further increase in wheel hardness giving optimum values. It appears that as the wheel hardness is increased the wheel first grinds in the self-sharpening region where bond-post fracture is predominant, later in a zone of mixed conditions—partly sharpening and partly blunting. In the final range, blunting dominates resulting in increased wheel loading and reduction in material removal rate. Similarly, when the grain size of the wheel is increased, the grinding ratio increases to reach a maximum value and then decreases as shown in Fig. 11.15(b). The curve is similar to that obtained with wheels of different grades and can be explained in a similar manner. Thus, for a given material there is an optimum grade and grain size that gives maximum grinding ratio and this wheel is the most suitable one.

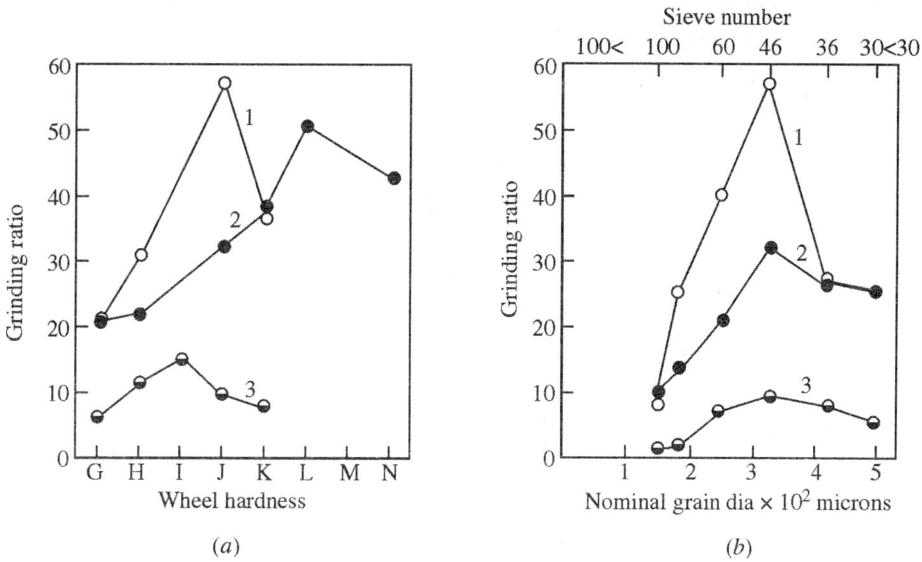

Fig. 11.15 Variation in grinding ratio with wheel hardness and grain size for different work materials (1) steel (17 R_c); (2) steel (29 R_c); (3) steel (17 R_B) (After Pandey *et al.*, 1980)

Referring to Fig. 11.11, it is obvious that for fine grinding, the wheel must be dressed and its cutting ability restored before the end of phase II. The life of the wheel thus obtained will be purely from the consideration of wheel wear, wheel loading and formation of wear flats. The other aspects that are often taken into consideration are surface finish, dimensional accuracy, grinding forces, grinding burn, vibration and noise. The choice of one or more of these parameters for establishing wheel life will obviously depend upon the technical requirements imposed. In view of the fact that large number of input variables affect the life of grinding wheel, it has not been possible to develop a simple equation for wheel life of the kind proposed for single-point cutting tools. Wheel life is not a consideration in stock removal operations where continous self-sharpening of the wheel occurs.

11.6 GRINDING TEMPERATURE

Consider again the chip-forming process in horizontal surface grinding depicted in Fig. 11.16. When the grain first contacts the workpiece surface, sliding and ploughing of the surface layer occurs. As the chip thickness increases, the forces increase and a layer of material is removed in the form of a chip. Heat is generated in the interference zone due to plastic deformation and frictional sliding. Further, heat is generated as the chip formed travels up the front face of the grain. Due to these, very high temperatures are recorded near the cutting edges. In addition to this temperature, the surface of the workpiece is subjected to the wheel-workpiece interference zone temperature.

The grinding zone temperature is, therefore, of interest from the point of view of wheel wear as well as the quality of workpiece produced. As a consequence of high temperature, not only the wheel wear increases but large residual tensile stress may develop at the surface layer of the workpiece resulting in surface cracks. Even micro-structural changes may occur if the temperature is sufficiently high. For example, during grinding of steel workpieces, the surface layer may become asutenite due to high temperature, and subsequent quenching by grinding fluid may transform it into martensite. A typical variation in residual stress distribution is shown in Fig. 11.17, which indicates that the stresses are confined to the top layer of the workpiece surface.

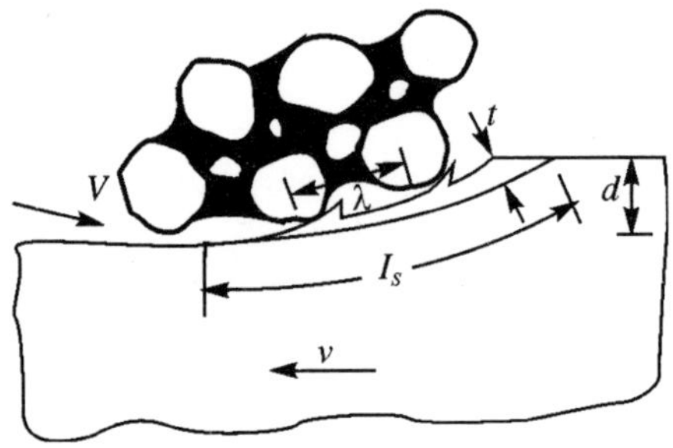

Fig 11.16 Chip formation prooooo

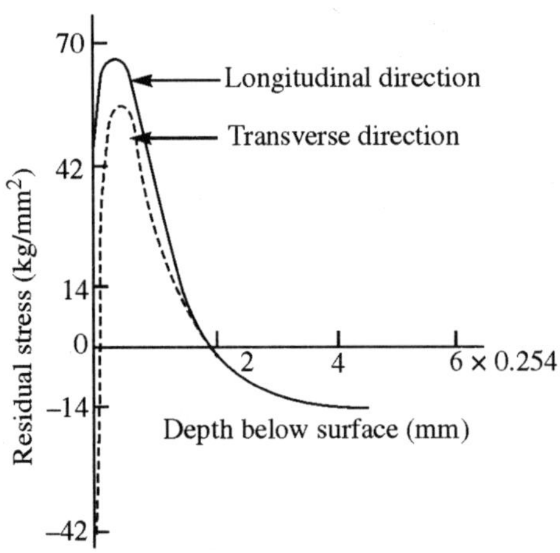

Fig. 11.17 A typical residual stress distribution (After Letner, 1955)

As indicated earlier, the main source of heat is in the grain-chip-workpiece interference zone. The length of this interference zone may be taken as the chip length l_s which is typically of the order of 1-3 mm in fine grinding. In the wheels used for fine grinding, the average spacing between successive grains is estimated to be in the range of 1-2 mm. Therefore, at a given instant one or two grains may be located along l_s in the plane of the figure. The surface speed of the wheel is, however, much higher than the work speed and many cutting points may pass over the contact length as a point on the workpiece surface passes under the wheel. Since each of these cutting points is a heat source, the workpiece may be assumed to be subjected to a continuously acting distributed heat source.

Jaegar's model for a continuously acting uniform band heat source on the surface of a moving semi-infinite body is illustrated in Fig. 11.18. It is assumed that there is no heat loss from the surface of the semi-infinite body. Jaeger's solution for temperature distribution θ at the surface ($z = 0$), of an infinite strip ($-l' < x < l', -\infty < y < \infty$), when the heat flux q and the surrounding medium moves with a velocity V' in the x-direction gives,

$$\theta = \frac{2\alpha_d q}{\pi K V'} \int_{X-L'}^{X+L'} e^{-u'} K_o(u')\, du' \qquad ...(11.24)$$

where $K_o\ (u')$ is the modified Bessel function of second kind of zero order, K and α_d are the thermal conductivity and thermal diffusivity, and

$$X = \frac{V'x}{2\alpha_d}\ ;\ L' = \frac{V'l'}{2\alpha_d} \quad \text{...(11.25)}$$

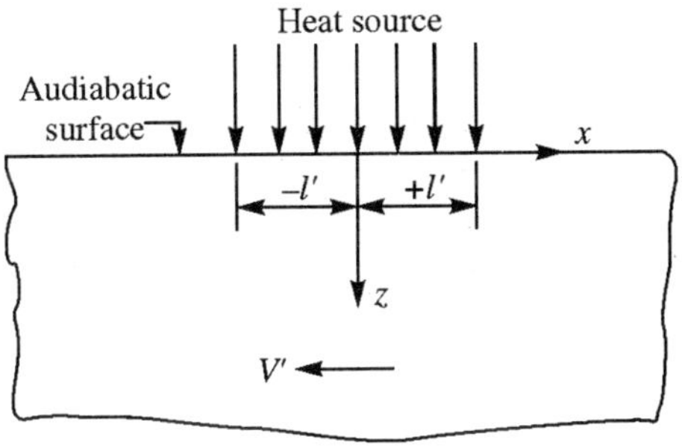

Fig. 11.18 Band heat source on semi-infinite body

When $L' > 5$, equation (11.24) can be approximated for obtaining the maximum surface temperature θ_m occurring at $x = -l'$ as

$$\theta_m = \frac{2ql'}{K\sqrt{(\pi L')}} \quad \text{...(11.26)}$$

and the average surface temperature under the heat source as

$$\bar{\theta} = \frac{4ql'}{3K\sqrt{(\pi L')}} = \frac{0.754ql'}{K\sqrt{L'}} \quad \text{...(11.27)}$$

Thus,

$$\bar{\theta} = \frac{2}{3}\theta_m \quad \text{...(11.28)}$$

Assuming that a constant and uniformly distributed heat source acts on the workpiece surface across the chip length during grinding, the variables V', l' and L' can be obtained as

$$V' = v$$

$$l' = \frac{l_s}{2} = \frac{1}{2}\sqrt{(dD)}$$

and

$$L' = \frac{l_s v}{2\alpha_d} \quad \text{...(11.29)}$$

Here the case of horizontal surface grinding (Fig. 11.2a) has been considered.

The strength of heat source q during surface grinding can be estimated as

$$q = \begin{bmatrix} \text{Fraction of total} \\ \text{grinding energy entering} \\ \text{the workpiece} \end{bmatrix} \times \begin{bmatrix} \text{The total grinding} \\ \text{energy per unit} \\ \text{surface area} \end{bmatrix}$$

or

$$\bar{q} = (0.8) \times \left(\frac{F_p V}{B l_s}\right) \qquad \text{...(11.30)}$$

for horizontal surface grinding.

In fine grinding operations, the width of the wheel in contact with the workpiece is much greater than the chip-length, *i.e.*, $B >> l_s$, and the approximation of an infinitely long band of heat source appears to be reasonable. If q is assumed to be uniformly distributed and invariant with time, then q may be replaced by $\bar{q}$ in equations (11.26) and (11.27). Experiments indicate that the fraction of heat that goes into the workpiece in fine grinding operations generally varies between 0.75 and 0.85 and its value may be taken as 0.8.

The maximum surface temperature θ_m and the average surface temperature $\bar{\theta}$ during surface grinding can now be estimated using equations (11.26) to (11.30). A similar approach may be adopted for estimating θ_m and $\bar{\theta}$ during external cylindrical and internal grinding.

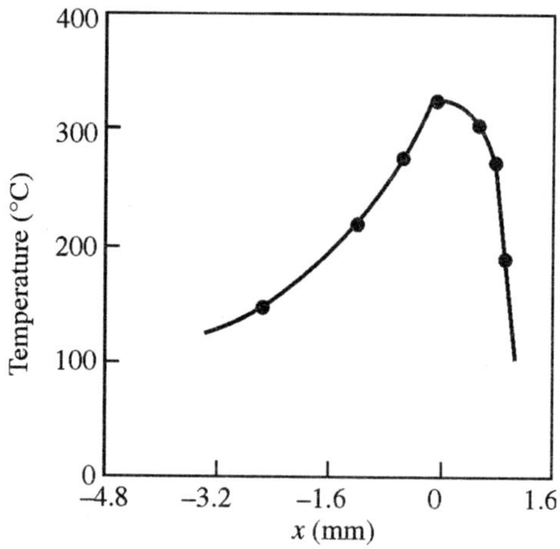

Fig. 11.19 Experimental temperature distribution on workpiece (After Sauer, 1972)

A typical temperature distribution on the surface of workpiece ($z = 0$) obtained experimentally using thermocouple arrangement is shown in Fig. 11.19. Here the maximum temperature (θ_m) of around 330°C occurs under the wheel ($x = 0$) at an instant of time and the average surface temperature ($\bar{\theta}$) will be around 200°C. The instantaneous temperature at the grain-workpiece interface is usually an order of magnitude higher (2000°C or more) but exists for a very short time to cause any significant damage to the ground surface. The wear properties of the wheel and the chemical interaction between the grain and workpiece may, however, get affected.

11.7 SURFACE ROUGHNESS

The surface produced on a horizontal surface grinding machine consists of scallops left on the workpiece due to finite spacing of successive grains. If all grains are assumed to be at the same radial distance and equally spaced, the plan view of the finished surface will consist of an array of slender-shaped scratches as shown in Fig. 11.20. Assuming the mid-point of the scratch to be the deepest point, the idealised mean peak-to-valley height h can be evaluated in the following way.

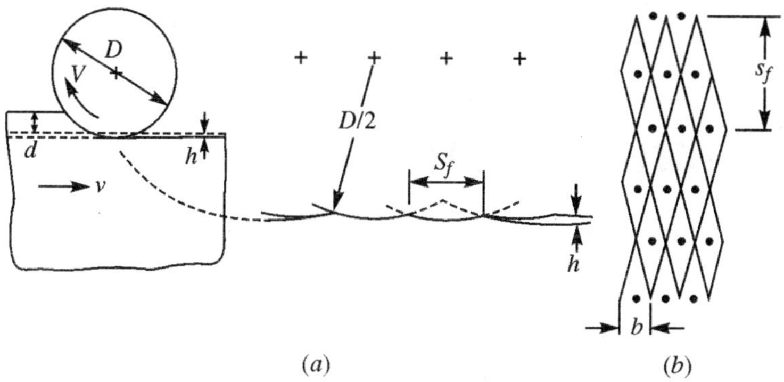

Fig. 11.20 Surface produced in horizontal surface grinding. (a) Scallops on the ground surface; (b) Plan view of scratches on the ground surface.

If there are C number of cutting points per unit area on the wheel surface and the average chip has a width b, then the distance between two successive cutting points (grain spacing) on the wheel surface will be $1/bC$. The corresponding distance on the workpiece surface will be v/V times this value. The distance, s_f, moved by the workpiece between passage of successive grains can be evaluated as

$$s_f = \frac{v}{VCb}$$

and using equation (11.15)

$$s_f = \frac{2v}{VCrt} \qquad \text{...(11.31)}$$

The distance s_f can also be evaluated in terms of h as

$$\left(\frac{s_f}{2}\right)^2 = \left(\frac{D}{2}\right)^2 - \left(\frac{D}{2} - h\right)^2$$

or

$$s_f = 2\sqrt{(Dh)} \qquad \text{...(11.32)}$$

Equating (11.31) and (11.32), we get

$$h = \left(\frac{2v}{VCrt}\right)^2 \frac{1}{4D}$$

Substituting for $\frac{v}{VCr}$ from equation (11.18), we get

$$h = \frac{t^2}{16d} \qquad \text{...(11.33)}$$

This shows that the chip thickness t is the most important variable affecting the surface finish. Therefore, for better finish, the grinding conditions should be chosen to reduce the value of t.

CHAPTER 12

Vibrations in Machining

12.1 INTRODUCTION

During machining operations, machine tools are subjected to static as well as dynamic loads associated with the cutting process and the movements of machine tool members. For example, during a turning operation some of the possible modes of vibration are:

(*a*) Bending of tool shank or tool post or compound slide.
(*b*) Bending of workpiece between centers.
(*c*) Workpiece or headstock in torsion.
(*d*) Bending of lathe bed.
(*e*) Bending of headstock or tailstock.

Since the point of application of load changes during the cutting process, the deflections also vary with time. These deflections may cause fluctuations in cutting speed, feed, depth of cut and inclination of tool faces.

During machining, dynamic forces also come into action and produce rapidly varying deflection in the machine-work-tool system. The main sources of these vibrations are:

(*a*) Generation of vibrations from the cutting process.
(*b*) Intermittent cutting.
(*c*) Inhomogeneties in the work material.
(*d*) Disturbances in the drive system.
(*e*) Transmission of vibrations from outside sources.

Both static and dynamic deflections cause dimensional and finish inaccuracies. Tool life also gets adversely affected under conditions of vibrations and may decrease by 25% under certain conditions. Undue vibrations also decrease the life of machine tool. If the frequency of vibration coincides with the natural frequency of any mode of the machine tool, serious damage may be caused to the machine tool.

There are features of cutting process which make it inherently periodic over certain cutting conditions. These include built-up-edge formations, chip segmentation and discontinuous chip formation. For a given cutting condition, the built-up-edge will grow to a critical size before the applied forces casuse its removal. The rate of growth of built-up-edge and its removal is nearly constant. Thus, there will be a natural frequency for built-up-edge formation.

Under certain cutting conditions, the plastic deformation zone is not constant but gradually spreads into the workpiece (Fig. 12.1) and then rapidly moves towards the tool. This sequence is repeated and with each oscillation of the plastic zone a chip segment is formed. This segmentation process in continuous chip causes periodic fluctuations in chip thickness. The frequency of fluctuations in the measured cutting force is generally found to correlate with the frequency of formation of chip segments.

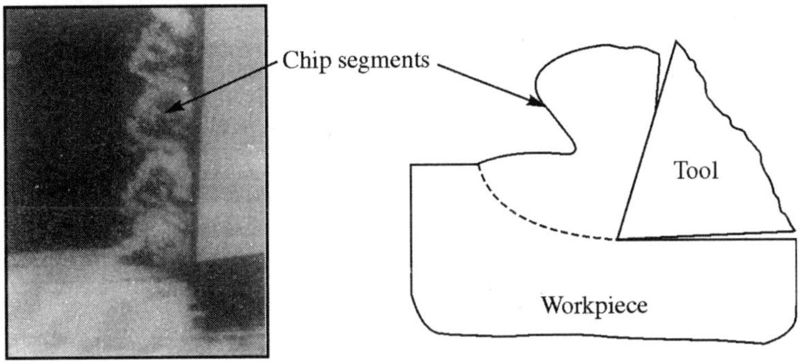

(a) (b)

Fig. 12.1 Segmental chips

In discontinuous chip formation also the chip segments are produced at a uniform rate with each segment going through a formation cycle at the end of which it gets ruptured from the parent body. The cyclic frequency in this case also is largely depended on the cutting conditions.

In intermittent cutting operations such as milling, the fluctuating cutting force generates vibration which gets transmitted to the machine tool. If this force fluctuation falls in the range of any natural frequency, severe vibration may be the result.

Inhomogeneties in the workpiece material, such as sudden increment in hardness, can generate an impulse force which may set-up transient vibration in the cutting tool and the machine tool.

Vibrations can be generated due to unbalanced rotating mass in the drive system or due to faulty arrangement of drive and supporting bearings. Vibrations from outside sources can also get transmitted to the machine tool from the shop floor.

12.2 MACHINING VIBRATIONS

Machining vibrations can be broadly classified into three categories:

(*a*) Free vibration

(*b*) Forced vibration

(*c*) Self-induced vibration

A *free vibration*, also called *transient vibration*, is caused by impact or shock condition. This can happen when the tool suddenly enters the workpiece or when an external impact get transmitted through the machine foundation. Such impacts cause a relatively large force to act for a short duration and as a consequence the machine tool starts to vibrate. Such vibrations, however, decay rapidly under the damping action of machine tool.

A *forced vibration* may get initiated in the cutting process or by sources other than cutting. The cutting process, under certain conditions, may become periodic and can cause forced vibrations. For example, periodicity during chip formation with built-up-edge, discontinuous chip formation and chip segmentation can lead to forced vibrations.

Forced vibration may also get initiated by unbalance in the machine-tool drive, misalignments in machine members, mounting defects, imperfections in bearings, or some dynamic load external to the machine. Vibrations due to these and also those due to the cutting process are normally not at the resonance frequency.

Under certain conditions, the machine tool can vibrate due to the cutting process itself. In such cases, the exciting force does not come from outside elements but is supplied by the cutting process itself. In these self-induced vibrations, the vibration movement causes the cutting process to provide energy to maintain the vibration. These types of self-induced vibration are commonly known as *machine-tool chatter.*

The self-excited and forced vibrations can be distinguished by the fact that self-excited vibrations are self-sustaining, drawing energy from its own periodic motion, while forced vibrations are controlled by the characteristics of the external alternating energy source. Thus, in a *self-excited vibration*, the motion stops when the alternating force disappears, while in a forced vibration the sustaining alternating force exists independently and persists even when the vibrating motion is stopped. The frequency of self-excited vibrations is, therefore, close to the natural frequency of the undamped system, while forced vibrations occur at the frequency of the alternating external energy source.

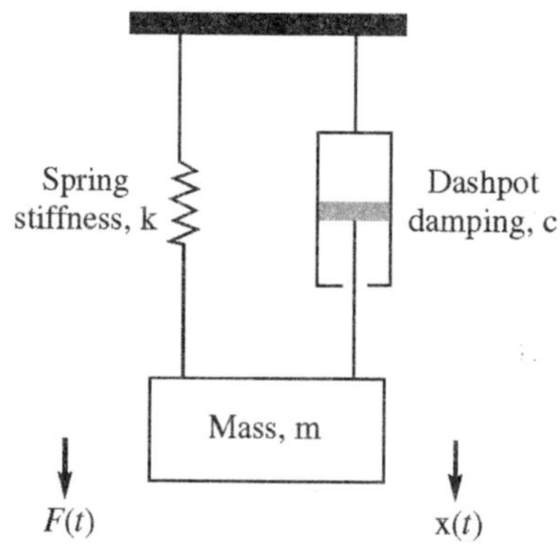

Fig. 12.2 A single-degree vibration system

Consider a single-degree of freedom system having a mass m on a restoring spring of stiffness k and a dashpot with damping c as shown in Fig. 12.2. When an external force $F(t)$ is applied to mass m, the equation of motion becomes

$$m\ddot{x} + c\dot{x} + kx = F(t) \qquad ...(12.1)$$

Here kx is the restoring force, $c\dot{x}$ is the damping force and $F(t)$ is the exciting force. The implied assumption here is that the damping force is proportional to $\dot{x}$. For free vibration, equation (12.1) reduces to

$$m\ddot{x} + c\dot{x} + kx = 0 \qquad ...(12.2)$$

When $c > 0$,

$$x = A \exp\ \{-\xi w_0\} t \cos\left\{\sqrt{(1-\xi^2)}\omega_0 t + \psi\right\} \qquad ...(12.3)$$

and its solution gives

$$x = A \exp\left(-\frac{c}{2m} + j\omega_0\right)t \qquad ...(12.4)$$

where $w_0 = \sqrt{k/m}$, $\xi = c/c_c$ and $c_c = 2\sqrt{km}$. Here A and ψ are constants and their values depend on displacement and velocity conditions at $t = 0$. Equation (12.4) shows that the amplitude of x decreases exponentially with time t (Fig. 12.3) and dies out because of positive damping.

When $c < 0$,

$$x = A \exp \{+\xi w_0 t\} \cos\left\{\sqrt{(1-\xi^2)}\, \omega_0 t + \psi\right\} \qquad ...(12.5)$$

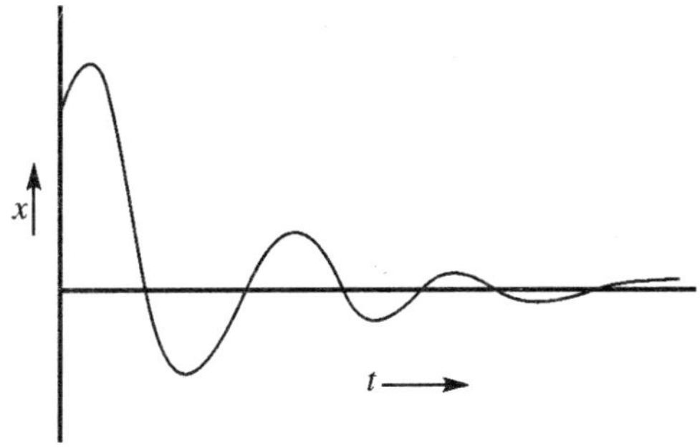

Fig. 12.3 Damped vibration causing the amplitude to decrease exponentially with time

This shows that the amplitude increases exponentially with time (Fig. 12.4) and the system beomes dynamically unstable. Such systems are susceptible to self-excited vibrations. $c = 0$ represents the threshold of stability when only free undamped vibration occurs.

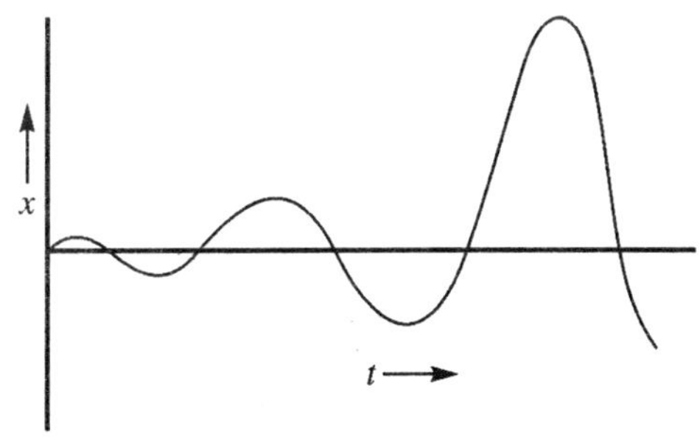

Fig. 12.4 Self-excitation causing the amplitude to increase exponentially with time

It is clear from the above analysis that a self-induced or self-excited vibration system can be represented by a negative term for damping, *i.e.*, a negative coefficient of $\dot{x}$ term.

If the system is subjected to an external force $F(t)$ in the form of $F'(t) + B\dot{x}$, the equation of motion becomes

$$m\ddot{x} + c\dot{x} + kx = F'(t) + B\dot{x} \qquad (12.6)$$

or

$$m\ddot{x} + (c - B)\dot{x} + kx = F'(t) \qquad (12.7)$$

and it is possible for coefficient of $\dot{x}$ to become negative and become an unstable system.

The above analysis indicates that in a dynamically unstable system, the amplitude should grow indefinitely. This, however, does not happen in practice primarily because with increasing amplitude the non-linearity of the system becomes increasingly prominent and growth of amplitude gets arrested. Therefore, in physical systems, self-excitation and damping co-exist and act separately.

The structure of machine tool has considerable influence on the dynamic stability of the system. With increased damping in the structure of machine tools, the dynamic stability of the system can be improved sufficiently to avoid chatter.

Free and forced vibrations are generally not much of a problem since they are normally of small amplitude. The factors that lead to cyclic forcing effects or self-induced vibration are of main interest as far as the cutting process is concerned.

12.3 CAUSES OF SELF-INDUCED VIBRATIONS

Two main causes of self-induced vibrations are *regenerative effect* and *mode-coupling effect*. Regeneration arises when the tool moves on surfaces it has machined previously, *i.e.*, when successive cuts overlap as shown in Fig. 12.5. If a vibration was present in the previous pass due to the variation in uncut chip thickness, the machined surface will be undulated (wavy) and cutting forces will fluctuate. In successive passes, the size of surface undulations may increase and lead to increase in vibration. In other words, the vibration is regenerated in successive passes. When the energy imparted by the fluctuating cutting force is sufficient to overcome the damping present in the system, the vibration continues to increase in magnitude. This self-excited vibration, commonly known as *regeneration chatter*, is not an effect, which is inherent to the cutting process but is a forcing effect which arises because a vibration was initiated.

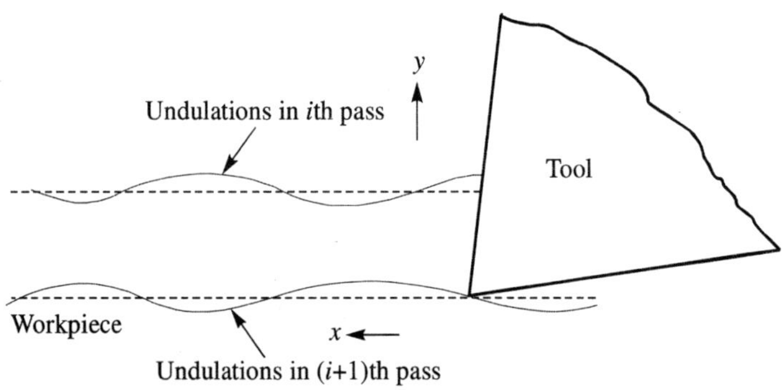

Fig. 12.5 Regenerative effect

If the amplitude of vibrations in *i*th pass is y_i, and that in $(i + 1)$th pass is $y_{(i+1)}$, the variation in amplitude during $(i + 1)$th pass is $[y_{(i+1)} - y_i]$. Since both y_i and $y_{(i+1)}$ are time-dependent vectors, their inter-relationship is generally expressed as

$$y_{(i+1)} = qy_i \exp(-j\phi) \qquad \text{...(12.8)}$$

where ϕ is the phase shift between successive undulations with $j = \sqrt{-1}$ and q is the amplification factor. Undulations, therefore, get created in each pass causing regeneration of vibration.

The mode-coupling effect occurs when the vibratory system possesses at least two degrees of freedom. In such a situation, a vibration initiated in any one direction may lead to variation in the cutting conditions. The resulting fluctuations in cutting force not only acts in the direction of vibration but also in the direction at right angle to this.

Consider a two degree of freedom system (Fig. 12.6) in which the system can vibrate simultaneously in x_1 and x_2 directions on x_1 O x_2 plane. The two paths, x_1 and x_2, are perpendicular to each other. For a given frequency of vibration ω, the difference in the amplitude of vibration in the two given directions and the difference in their relative phase results in a loop motion of cutting edge as shown in the figure. This looping of cutting edge causes the cutting force F to oppose the motion of the vibrating mass in the first half and assist it during the other half. The energy of the system, therefore, diminishes in the first half and increases in the second half.

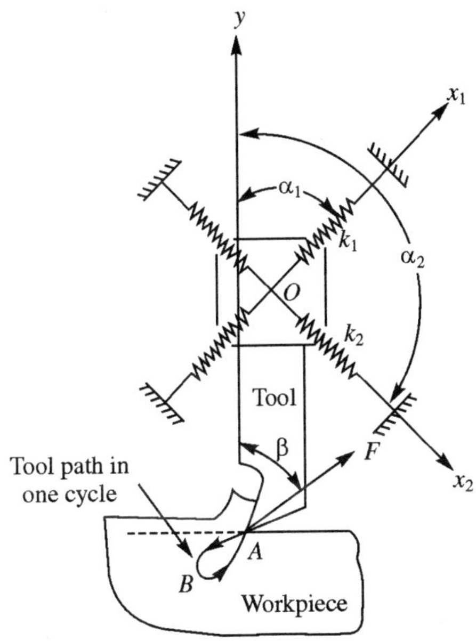

Fig. 12.6 Mode coupling effect

The loop motion causes the uncut chip thickness to be greater during the second half and, therefore, the cutting force is higher. Because of this, the addition of energy during this period is more than its expenditure. Thus, there is a net input of energy to the vibratory system. This may be sufficient to overcome the damping present in the system and cause the amplitude of vibration to increase during each cycle.

12.4 ANALYSIS OF SELF-EXCITED VIBRATION

A number of analyses are available to derive the limit of stability for the machine tool and cutting system. One of the simple, yet useful, single-degree of freedom model is due to Tlusty and Polacek (1963). This analysis considers the effects of regeneration as well as mode coupling.

12.4.1 Single-degree-of-freedom Analysis with Regenerative Effect

Tlusty and Polacek model assumes that F (Fig. 12.7) is the only external force acting on the system and any increment to force F due to regeneration effect occurring in y-direction continues to act along $F(t)$ direction.

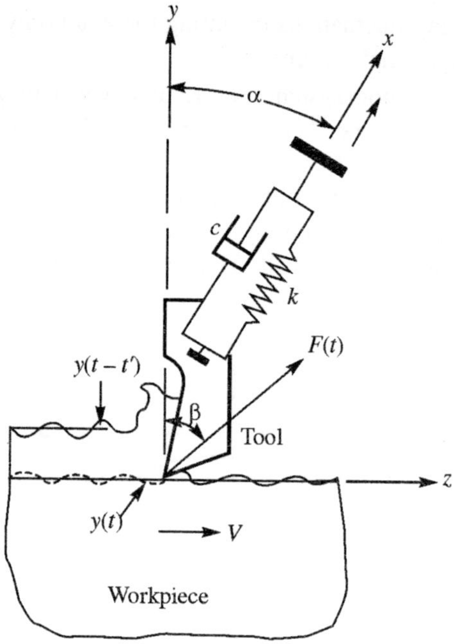

Fig. 12.7 Regeneration effect

If the principal mode (x) is inclined at an angle α to y-direction, the motion along y and x are related as

$$y = x \cos \alpha \qquad \text{...(12.9)}$$

At any instant t, the chip thickness is the difference between the surface generated during the previous cut and the instantaneous position of tool as shown in Fig. 12.7. If the corresponding point on the machine surface was machined at time $(t - t')$, the chip thickness variation can be estimated as

$$y = y(t) - y(t - t')$$

or

$$y = [x(t) - x(t - t')]\cos\alpha \qquad \text{...(12.10)}$$

If γ is the *coupling coefficient,* assumed real and positive, between force $F(t)$ along y and z, then

$$F(t) = -\gamma y$$

or

$$F(t) = -r[x(t) - x(t - t')]\cos\alpha \qquad \text{...(12.11)}$$

The equation of motion along x-axis can be written as

$$m\ddot{x} + c\dot{x} + kx = F_x(t) = F(t)\cos(\alpha - \beta) \qquad \text{...(12.12)}$$

and its solution (using complex exponential notation e^{jwt} for harmonic quantities with $j = \sqrt{-1}$) gives,

$$x(t) = \frac{F(t)\cos(\alpha - \beta)}{k\left[1 - \left(\frac{\omega}{\omega_0}\right)^2 + j2\xi\left(\frac{\omega}{\omega_0}\right)\right]} \qquad \text{...(12.13)}$$

where ω is the forcing frequency, ξ is the damping factor (= c/c_c), $\omega_0 = (k/m)^{1/2}$ and $r = \omega/\omega_0$.

Substituting of $F(t)$ from equation (12.11) and rearranging, gives

$$x(t)\left[\frac{k}{\cos\alpha\cos(\alpha-\beta)}\right][1-r^2+j2\xi r] = -\gamma(x(t)-x(t-t')] \qquad ...(12.14]$$

Defining ϕ as

$$\phi = \frac{\cos\alpha\cos(\alpha-\beta)}{k\{(1-r^2)+j2\xi r\}} \qquad ...(12.15)$$

equation (12.14) becomes

$$\frac{x(t)}{\phi} = -\gamma[x(t)-x(t-t')] \qquad ...(12.16)$$

Hence,

$$\frac{x(t)}{x(t-t')} = \frac{\gamma}{\left(\frac{1}{\phi}+\gamma\right)} = \frac{\phi}{\left(\phi+\frac{1}{\gamma}\right)} = q \qquad ...(12.17)$$

This equation indicates that for the system to be stable q must be less than 1, *i.e.*, when $q < 1$ the amplitude does not build-up. When $q > 1$ the system is unstable and $q = 1$ defines the threshold of stability.

For real r,

$$\phi = \frac{\cos\alpha\cos(\alpha-\beta)}{k}\left[\frac{1-r^2}{(1-r^2)^2+4\xi^2r^2} - j\frac{2\xi r}{(1-r^2)^2+4\xi^2r^2}\right] \qquad ...(12.18)$$

or

$$\phi = G + jH \qquad ...(12.19)$$

and for real ϕ, equation (12.17) gives

$$q = 1 = \frac{|G|}{\left(|G|+\frac{1}{\gamma}\right)} \qquad ...(12.20)$$

for threshold of stability.

Equation (12.20) is satisfied when

$$-|G| = |G| + \frac{1}{\gamma} \qquad ...(12.21)$$

and

$$\gamma = -\left(\frac{1}{2G}\right) \qquad ...(12.22)$$

gives the stability condition.

When $\alpha - \beta = 0$, the *limiting coefficient of coupling* can be defined by

$$\gamma^*_{max} = -\left(\frac{1}{2G_{min}}\right) \quad ...(12.23)$$

12.5 TWO-DEGREE-OF-FREEDOM ANALYSIS WITH MODE COUPLING EFFECTS

In the analysis presented above only the regenerative effect was considered. It has been established by several researchers that instability may arise in the machine-tool and cutting system even in the absence of regeneration effect through mode coupling. In order to understand this, a two-degree-of-freedom system with loop motion of the cutting edge (Fig. 12.6) is considered. This analysis was also developed by Tlusty and Polacek (1963).

Figure 12.6 indicates that during the loop motion from A to B, the motion is opposed to force F which causes dissipation of energy. During loop motion from B to A the energy input increases because of increased depth of cut. The net effect can produce self-excited vibration. Expressing the increase in force F due to motion along y-direction as $dF = -\gamma y$, where γ is the coupling coefficient, the equations of motion of the mass system having two springs at right angles as shown in Fig. 12.6 can be written as

$$\begin{aligned} m\ddot{x}_1 + k_1x_1 &= dF\cos(\beta - \alpha_1) = -\gamma y\cos(\beta - \alpha_1) \\ m\ddot{x}_2 + k_2x_2 &= dF\cos(\beta - \alpha_2) = -\gamma y\cos(\beta - \alpha_2) \end{aligned} \quad ...(12.24)$$

Now

$$y = x_1\cos\alpha_1 - x_2\sin\alpha_1 = x_1\cos\alpha_1 + x_2\cos\alpha_2 \quad ...(12.25)$$

since

$$\alpha_1 = \alpha_2 - 90° \quad ...(12.26)$$

Equations (12.24) now become

$$\begin{aligned} & m\ddot{x}_1 + [k_1 + \gamma\cos\alpha_1\cos(\beta - \alpha_1)]x_1 + [\gamma\cos\alpha_2\cos(\beta - \alpha_1)]x_2 = 0 \\ & m\ddot{x}_2 + [k_2 + \gamma\cos\alpha_2\cos(\beta - \alpha_2)]x_2 + [\gamma\cos\alpha_1\cos(\beta - \alpha_2)]x_1 = 0 \end{aligned} \quad ...(12.27)$$

The last terms in the above equations represent the mode coupling effect.

Assuming the solution of the above equations to be of the form,

$$x_1 = A_1e^{jwt} \text{ and } x_2 = A_2e^{jwt} \quad ...(12.28)$$

equations (12.27) become

$$\begin{aligned} & A_1(k_1 + \gamma u_1 - m\omega^2) + A_2\cos\alpha_2\cos(\alpha_1 - \beta) = 0 \\ & A_1\gamma\cos\alpha_1\cos(\alpha_2 - \beta)A_2(k_2 + mu_1 - m\omega^2) = 0 \end{aligned} \quad ...(12.29)$$

where

$$u_1 = \cos\alpha_1\cos(\beta - \alpha_1); \quad u_2 = \cos\alpha_2\cos(\beta - \alpha_2) \quad ...(12.30)$$

From equations (12.29), the frequency equation can be obtained as

$$\Delta\omega = \begin{vmatrix} k_1 + \gamma u_1 - m\omega^2 & \cos\alpha_1\cos(\alpha_1 - \beta) \\ \gamma\cos\alpha_2\cos(\alpha_2 - \beta) & k_2 + \gamma u_2 - m\omega^2 \end{vmatrix} = 0 \quad ...(12.31)$$

which on simplification gives

$$\frac{1}{\gamma}+\frac{u_1}{k_1-m\omega^2}+\frac{u_2}{k_2-m\omega^2}=0 \qquad \text{...(12.32)}$$

For a decoupled system,

$$m=\frac{k_1}{\omega_1^2}=\frac{k_2}{\omega_2^2} \qquad \text{...(12.33)}$$

Therefore, equation (12.32) can be rewritten as

$$\frac{1}{\gamma}=\sum_{i=1}^{2}\phi_i=0 \qquad \text{...(12.34)}$$

where

$$\phi_i=\frac{u_i}{k_i(1-\gamma_i^2)} \qquad \text{...(12.35)}$$

and

$$\gamma_i=\frac{\omega}{\omega_i} \qquad \text{...(12.36)}$$

The stability criterion for limiting γ can now be defined as

$$\gamma^*=\frac{1}{\sum\phi_i} \qquad \text{...(12.37)}$$

The above analyses indicate that self-excited vibration may be prevented by appropriately changing the value of γ. This can be achieved by changing the cutting conditions (speed, feed and depth of cut) and the tool geometry. Vibration can also be prevented by changing the characteristics of the vibratory system through method of clamping the workpiece, stiffening of offending modes, and use of additional damping.

Numerical Examples

Example 1. A straight turning tool has a back rake angle α_b of 8° and a side cutting edge angle γ_s of 30°. What value of side rake angle α_s should be taken for achieving orthogonal cutting condition?

Solution:

For orthogonal cutting, $i = 0$. Therefore equation (3.1) gives

$$\tan i = \sin\gamma_p \tan\alpha_b - \cos\gamma_p \tan\alpha_s = 0$$

In the present problem

$$\alpha_b = 8^\circ$$

and

$$\gamma_p = 90^\circ - \gamma_s = 60^\circ$$

Therefore,

$$\tan\alpha_s = \frac{\sin 60^\circ \tan 8^\circ}{\cos 60^\circ}$$

$$= \tan 60^\circ \tan 8^\circ = 0.24$$

and

$$\alpha_s = 13.5^\circ$$

Example 2. What are the back rake and side rake angles on a 0-10-6-6-8-75-1 (mm) ORS cutting tool?

Solution:

For this tool,

$$i = 0^\circ, \alpha_o = 10^\circ \text{ and } \gamma_p = 75^\circ$$

Equations (3.1) and 3.2) give

$$\tan i = \sin\gamma_p \tan\alpha_b - \cos\gamma_p \tan\alpha_s$$

and

$$\tan \alpha_o = \sin\gamma_p \tan\alpha_s + \cos\gamma_p \tan\alpha_b$$

Substituting for i, α_o and γ_p, we get

$$\sin 75° \tan \alpha_b - \cos 75° \tan \alpha_s = \tan 0°$$

and

$$\cos 75° \tan \alpha_b + \sin 75° \tan \alpha_s = \tan 10°$$

or

$$-(\sin^2 75° + \cos^2 75°) \tan \alpha_s = -\sin 75° \tan 10°$$

or

$$\tan \alpha_s = 0.170$$
$$\alpha_s = 9.65°$$

Similarly,

$$(\sin^2 75° + \cos^2 75°) \tan \alpha_b = \cos 75° \tan 10°$$

or

$$\tan \alpha_b = 0.0456$$
$$\alpha_b = 2.61°$$

Example 3. In a conventional twist drill, the point and helix angles are 210° and 30°, respectively. If the chisel edge diameter is 4 mm and the drill diameter is 25 mm, calculate the minimum and maximum normal rake angle α_n at the lips.

Solution:

The normal rake angle of a twist drill can be evaluated from equation (10.42) which gives

$$\tan \alpha_n = \frac{r \tan \psi}{\frac{D}{2} \sin \frac{\beta}{2}}$$

In this problem,

$$\psi = 30°$$
$$\beta = 120°$$
$$D = 25 \text{ mm}$$

and r varies from 2 mm to 12.5 mm. α_n is minimum when r = 2 mm and maximum when r = 12.5 mm. Thus,

$$(\alpha_n)_{\min} = \tan^{-1}\left[\frac{2 \tan 30°}{12.5 \sin 60°}\right] = \tan^{-1} 0.107 = 6.08°$$

and

$$(\alpha_n)_{\max} = \tan^{-1}\left[\frac{12.5 \tan 30°}{12.5 \sin 60°}\right] = \tan^{-1} 0.666 = 33.66°$$

Example 4. Results obtained during orthogonal machining of a steel with a single-point cutting tool of rake angle 10° are

$$t_c = 1.5 \text{ mm}, F_p = 980 \text{ N and } F_q = 440 \text{ N}$$

when t was 0.6 mm and b was 1.8 mm. Determine the friction force F at the chip-tool interface and shear strength of the material k.

Solution:

Equation (4.12) gives friction force F as

$$F = F_p \sin\alpha + F_q \cos\alpha$$

$$= 980 \sin 10^\circ + 440 \cos 10^\circ$$

$$= 603.49 \text{ N}$$

From equation (4.25),

$$r = \frac{t}{t_c} = \frac{0.6}{1.5} = 0.4$$

Using equation (4.28),

$$\tan\phi = \frac{r\cos\alpha}{1 - r\sin\alpha} = \frac{0.4\cos 10^\circ}{1 - 0.4\sin 10^\circ} = 0.423$$

or

$$\phi = 22.94^\circ$$

From equation (4.3),

$$A_s = \frac{bt}{\sin\phi} = \frac{1.8 \times 0.6}{\sin 22.94^\circ} = 2.77 \text{mm}^2$$

The shear force component F_s from equation (4.14) is

$$F_s = F_p \cos\phi - F_q \sin\phi$$

$$= 980 \cos 22.94^\circ - 440 \sin 22.94^\circ = 731 \text{ N}$$

Finally, the shear strength of the material k is obtained from

$$k = \frac{F_s}{A_s} = \frac{731}{2.77} = 263.9 \text{ N/mm}^2$$

Example 5. During orthogonal machining with an HSS tool, the rake angle was 5°, the undeformed chip-thickness was 0.25 mm and the width of cut was 4 mm. Taking the shear strength of the work material to be 350 N/mm^2 and the coefficient of friction between the chip and the tool to be 0.5, estimate the cutting force F_p and the thrust force F_q.

Solution:

Since the shear angle ϕ is not known, we shall estimate it from the Merchant's shear angle relationship (equation 4.9), *i.e.*,

$$\phi = \frac{\pi}{4} - \frac{1}{2}(\lambda - \alpha)$$

In this problem,

$$\lambda = \tan^{-1}(0.5) = 26.57^\circ$$

and

$$\alpha = 5^\circ$$

Thus,

$$\phi = 45^\circ - \frac{(26.57^\circ - 5^\circ)}{2} = 34.22^\circ$$

Equation (4.6) gives

$$F_p = \frac{k\,bt\cos(\lambda-\alpha)}{\sin\phi\cos(\phi+\lambda-\alpha)}$$

$$= \frac{350\times 4\times 0.25\cos(26.57^\circ - 5^\circ)}{\sin(34.22^\circ)\cos(34.22^\circ + 26.57^\circ - 5^\circ)} = 1029.44 \text{ N}$$

and using equation (4.7), F_q can be calculated as

$$F_q = \frac{k\,bt\sin(\lambda-\alpha)}{\sin\phi\cos(\phi+\lambda-\alpha)}$$

$$= \frac{350\times 4\times 0.25\sin(26.57^\circ - 5^\circ)}{\sin(34.22^\circ)\cos(34.22^\circ + 26.57^\circ - 5^\circ)} = 406.96 \text{ N}$$

F_p and F_q can also be evaluated from equations (4.10) and (4.11) since ϕ has been evaluated using equation (4.9). Thus,

$$F_p = 2\,k\,bt\cot\phi = 2\times 350\times 4\times 0.25\cot 34.22^\circ = 1029.25 \text{ N}$$

and

$$F_q = k\,bt\,(\cot^2\phi - 1) = 350\times 4\times 0.25\,(\cot^2 34.22^\circ - 1) = 406.68 \text{ N}$$

Example 6. During conventional shaping of a steel block with a 0-10-7-7-10-75-1.0 mm (ORS) HSS tool at a depth of cut of 3.5 mm and a feed of 0.26 mm/stroke, the chip-thickness was found to be 0.45 mm. If the shear strength of work material is 410 N/mm^2 and the coefficient of friction at the tool-chip interface is 0.6, estimate the force components F_p in the direction of cutting velocity, F_q in the direction of feed motion and F_r in the direction perpendicular to the surface generated.

Solution:

The undeformed chip-thickness $t = f\cos\gamma_s = 0.26\sin 75^\circ = 0.25$mm. from equation (10.22). Using equation (4.25),

$$r = \frac{t}{t_c} = \frac{0.25}{0.45} = 0.556$$

Using equation (4.28),

$$\tan\phi = \frac{r\cos\alpha}{1 - r\sin\alpha} = \frac{0.556\cos 10^\circ}{1 - 0.556\sin 10^\circ} = 0.61$$

Hence,

$$\phi = 31.38^\circ$$

The width of cut $b = d/\cos\gamma_s = 3.5/\sin 75^\circ = 3.62$ mm from equation (10.23). In the present case,

$$\lambda = \tan^{-1}(0.6) = 31^\circ$$

Since $i = 0$, we can use equations (4.6) and (4.7) for evaluating F_p and F_q components as shown in Fig. 10.8. Thus,

$$F_p = \frac{kbt\cos(\lambda-\alpha)}{\sin\phi\cos(\phi+\lambda-\alpha)}$$

$$= \frac{410\times3.62\times0.25\cos(31^\circ-10^\circ)}{\sin31.38^\circ\cos(31.38^\circ+31^\circ-10^\circ)} = 1089.8 \text{ N}$$

and

$$F_q' = \frac{kbt\sin(\lambda-\alpha)}{\sin\phi\cos(\phi+\lambda-\alpha)}$$

$$= \frac{410\times3.62\times0.25\sin(31^\circ-10^\circ)}{\sin(31.38^\circ)\cos(31.38^\circ+31^\circ-10^\circ)}$$

$$= 418.34 \text{ N}$$

Using Fig. 10.8, the feed component $F_q = F_q' \sin 75^\circ = 405.06$ N and the normal component $F_r = F_q' \cos 75^\circ = 108.53$ N.

Example 7. During orthogonal cutting with an HSS tool of specification 0-15-6-6-8-60-1 mm (NRS) on a mild steel tube at a cutting speed of 150 m/min and a feed rate of 0.4 mm/rev. The thickness of the chip was measured to be 0.8 mm. Calculate the shear strain undergone by the chip. If the thickness of the shear zone is measured as 9 microns, what is the rate of strain?

Solution:

Using equation (10.2),

$$t = f\cos\gamma_s = f\sin\gamma_p$$

$$= 0.4 \sin 60^\circ$$

$$= 0.346 \text{ mm}$$

Now

$$r = \frac{t}{t_c} = \frac{0.346}{0.8} = 0.43$$

and the shear angle, ϕ, can be evaluated from

$$\tan\phi = \frac{r\cos\alpha}{1-r\sin\alpha}$$

Therefore,

$$\phi = \tan^{-1}\left[\frac{0.43\cos15^\circ}{1-0.43\sin15^\circ}\right] = \tan^{-1}0.467 = 25^\circ$$

The shear strain γ can now be evaluated using equation (4.19), *i.e.*,

$$\gamma = \tan(\phi-\alpha)+\cot\phi$$

$$= \tan(25^\circ-15^\circ)+\cot25^\circ = 2.321$$

The shear strain rate $\dot{\gamma}$ can be estimated using equation (4.22). Thus,

$$\dot{\gamma} = \frac{\cos\alpha}{\cos(\phi-\alpha)}.\frac{V}{\Delta y}$$

$$= \frac{\cos 15^\circ}{\cos(25^\circ - 15^\circ)} \times \frac{150/60}{9\times10^{-6}}$$

$$= 2.7 \times 10^5 \text{ sec}^{-1}$$

Example 8. During machining of a steel with 7-α_s-6-6-8-30-1mm (ASA) HSS tool the undeformed chip-thickness of 0.18 mm and a width of cut of 2.5 mm were used. The side rake angle α_s of the tool was so chosen that the machining operation could be approximated to be orthogonal cutting. The force components measured using a two-component dynamometer were

Tangential cutting force	$F_P = 1177$ N
Thrust force	$F_q = 560$ N

Calculate (*i*) the side rake angle α_s, (*ii*) the coefficient of friction on the rake face and (*iii*) the dynamic shear strength of the work material *k*.

Solution:

(*i*) For orthogonal machining, the inclination angle $i = 0$. Using equation (3.1), we get

$$\tan\alpha_s = \frac{\sin\gamma_p \tan\alpha_b}{\cos\gamma_p}$$

where $\gamma_p = 90 - \gamma_s = 90 - 30 = 60^\circ$

Hence,

$$\tan\alpha_s = \tan 60^\circ \tan 7^\circ = 0.213$$

or

$$\alpha_s = 12^\circ$$

(*ii*) Equation (3.2) gives

$$\tan\alpha_o = \sin\gamma_p \tan\alpha_s + \cos\gamma_p \tan\alpha_b$$

$$= \sin 60^\circ \tan 12^\circ + \cos 60^\circ \tan 7^\circ = 0.245$$

Hence,

$$\alpha_o = 13.8^\circ$$

Using equation (4.16) we can directly determine the coefficient of friction as

$$\mu = \frac{F_q + F_p \tan\alpha}{F_p - F_q \tan\alpha} = \frac{560 + 1177 \tan 13.8^\circ}{1177 - 560 \tan 13.8^\circ} = 0.817$$

Here $\alpha = \alpha_o = \alpha_n$ since $i = 0$.

(*iii*) Since ϕ is not known; we shall evaluate it from Merchant's shear angle relationship. Thus,

$$\phi = \frac{\pi}{4} - \frac{1}{2}(\lambda - \alpha)$$

$$= \frac{\pi}{4} - \frac{1}{2}(\tan^{-1} 0.817 - 13.8°)$$

$$= 45° - \frac{1}{2}(39.25° - 13.8°) = 32.28°$$

The shear force can be evaluated as

$$F_s = F_p \cos\phi - F_q \sin\phi$$

$$= 1177 \cos 32.28° - 560 \sin 32.28° = 696.0 \text{ N}$$

using equation (4.14).

The dynamic shear strength

$$k = \frac{F_s \sin\phi}{bt}$$

from equation (4.17). Thus,

$$k = \frac{696 \times \sin 32.28°}{2.5 \times 10^{-3} \times 0.18 \times 10^{-3}} = 826.0 \times 10^6 \text{ N/m}^2$$

Example 9. During orthogonal cutting with a 0° – (–5°) –3° – 7° –10° –60° – 2.5 mm (ORS) carbide tool, the undeformed chip-thickness and the width of cut were fixed at 1.5 mm and 2.5 mm, respectively. The shear strength of the work material was 450 N/mm^2 and the rake face coefficient of friction was 0.6. The shear angle was evaluated to be 37° from experiments. Estimate the cutting and the thrust force components.

Solution:

In the present case,

$$t = 1.5 \text{ mm}, b = 2.5 \text{ mm}, k = 450 \text{ N/mm}^2, \phi = 37°$$

and

$$\lambda = \tan^{-1}(0.6) = 31°$$

From the specification of tool,

$$i = 0 \text{ and } \alpha_s = -5°$$

Therefore,

$$\alpha = \alpha_n = \alpha_o = -5°$$

Substituting these in equations (4.6) and (4.7), we get the cutting force,

$$F_p = \frac{k\,bt\cos(\lambda - \alpha)}{\sin\phi\cos(\phi + \lambda - \alpha)}$$

$$= \frac{450 \times 2.5 \times 1.5 \cos(31° + 5°)}{\sin 37° \cos(37° + 31° + 5°)} = 7758.95 \text{ N}$$

and the thrust force

$$F_q = \frac{k\,bt\sin(\lambda - \alpha)}{\sin\phi\cos(\phi + \lambda - \alpha)}$$

$$= \frac{450 \times 2.5 \times 1.5 \sin(31° + 5°)}{\sin 37° \cos(37° + 31° + 5°)} = 5637.2 \text{ N}$$

Example 10. During turning of a 100 mm diameter × 2 mm thick aluminium tube with a 0–15–8–8–12–75–1 mm (ORS) tool, the following data were recorded:

Cutting speed	:	150 m/min
Feed	:	0.2 mm/rev.
Cutting force	:	1600 N
Thrust force	:	850 N
Chip thickness	:	0.40 mm

Evaluate the shear angle and the power required for the cut.

Solution:

Equation (4.28) gives

$$\tan\phi = \frac{r\cos\alpha}{1 - r\sin\alpha}$$

In the present problem, $i = 0$, therefore,

$$\alpha = \alpha_n = \alpha_0 = 15°$$

The undeformed chip-thickness in turning is given by equation (10.2) as

$$t = f\cos\gamma_s = f\sin\gamma_p = 0.2 \sin 75° = 0.193 \text{ mm}$$

Therefore,

$$r = \frac{t}{t_c} = \frac{0.193}{0.40} = 0.48$$

and the shear angle,

$$\phi = \tan^{-1}\left[\frac{0.48\cos 15°}{1 - 0.48\sin 15°}\right]$$

$$= \tan^{-1} 0.529 = 27.9°$$

Power required P_w can be evaluated from

$$P_w = F_p V = 1600 \times 150$$

$$= 2.4 \times 105 \text{ Nm/min}$$

Example 11. During orthogonal cutting, with a 10° rake angle tool, assuming idealized stress distribution on the rake face, the following data were obtained.:

$h = 0.75$ mm $\qquad h' = 0.5$ mm

$\tau = 70 \times 10^6$ N/m^2 $\qquad \sigma = 200 \times 10^6$ N/m^2

Using Oxley's analysis calculate the average coefficient of friction along the rake face and the magnitude and direction of the resultant force for a width of cut $b = 6$ mm.

Solution:

For the idealized stress distribution shown in Fig. 4.23,

$$F = \frac{\tau}{2}(h + h')b = \frac{70\times10^6}{2}\times\frac{1.25\times6}{10^6} = 262.5 \text{ N}$$

and

$$N = \frac{\sigma}{2}hb = \frac{200\times10^6}{2}\times\frac{0.75\times6}{10^6} = 450 \text{ N}$$

Therefore,

$$\mu = \frac{F}{N} = \frac{262.5}{450} = 0.583$$

and

$$\lambda = \tan^{-1} 0.583 = 30°15'$$

The resultant force,

$$R = \frac{N}{\cos\lambda} = \frac{450}{0.864} = 520.8 \text{ N}$$

or

$$R = \frac{F}{\sin\lambda} = \frac{262.5}{0.504} = 520.8 \text{ N}$$

The resultant force of 520.8 N acts at an angle of 30° 15' from the normal to the rake face or (30° 15'– 10°) = 20° 15′ from the cutting direction.

Example 12. Using Oxley's approximate shear angle relationship, evaluate the power requirement when $\alpha = 20°$, $\tan\lambda = 1$, $t = 0.5$ mm, $b = 3$ mm, $V = 80$ m/min, $k = 450$ N/mm^2.

Solution:

Oxley's approximate shear angle relationship gives

$$\phi = 50° - 0.8(\lambda - \alpha)$$

or

$$\phi = 50° - 0.8\ (45° - 20°) = 30°$$

since $\lambda = \tan^{-1} 1 = 45°$

Power component of force,

$$F_P = \frac{ktb}{\sin\phi} \cdot \frac{\cos(\lambda - \alpha)}{\cos\theta}$$

$$= \frac{450 \times 0.5 \times 3 \cos 25°}{\sin 30° \cos 55°} = \frac{450 \times 0.5 \times 3 \times 0.9}{0.5 \times 0.57} = 2131.6 \text{ N}$$

since

$$\theta = \phi + \lambda - \alpha = 30° + 45° – 20° = 55°$$

Power required,

$$P_w = F_P V = 2131.6 \times 80 = 1.7 \times 10^5 \text{ Nm/min}$$

Example 13. During an oblique cutting test with $\alpha_n = 20°$, $i = 20°$, $t = 0.1$ mm, and $b = 3$ mm the following data were recorded:

Chip-thickness ratio $r_t = 0.42$

Force components: $F_P = 750$ N, $F_Q = 420$ N and $F_R = 180$ N

Evaluate the normal shear angle, the shear stress and the coefficient of friction assuming Stabler's chip-flow rule.

Solution:

Normal shear angle can be evaluated from

$$\tan\phi_n = \frac{r_t \cos\alpha_n}{1-r_t \sin\alpha_n} = \frac{0.42\cos 20°}{1-0.42\sin 20°} = \frac{0.42\times 0.94}{1-0.42\times 0.34} = 0.46$$

Therefore,

$$\phi_n = 24° \; 42'$$

Shear stress,

$$\tau = \frac{F_s}{A_s}$$

$$= \frac{\sqrt{\{(F_P \cos i + F_R \sin i)\cos\phi_n - F_Q \sin\phi_n\}^2 + (F_P \sin i - F_R \cos i)^2}}{bt/\cos i \sin\phi_n}$$

$$= \frac{\sqrt{\{(750\times 0.94 + 180\times 0.34)0.908 - 420\times 0.418\}^2 + (750\times 0.34 - 180\times 0.94)^2}}{3\times 0.1/0.94\times 0.418}$$

$$= \frac{527.17}{0.76} = 693.65 \text{ N/mm}^2$$

Coefficient of friction,

$$\mu = \frac{F}{N}$$

$$= \frac{\sqrt{\{(F_P \cos i + F_R \sin i)\sin\alpha_n + F_Q \cos\alpha_n\}^2 + (F_P \sin i - F_R \cos i)^2}}{(F_P \cos i + F_R \sin i)\cos\alpha_n - F_Q \sin\alpha_n}$$

$$= \frac{\sqrt{\{(750\times 0.94 + 180\times 0.34)0.34 + 420\times 0.94\}^2 + (750\times 0.34 - 180\times 0.94)^2}}{(750\times 0.94 + 180\times 0.34)0.94 - 420\times 0.34}$$

$$= \frac{660.90}{577.4} = 1.144$$

Example 14. During an oblique cutting test, the following data were recorded: α_n = 20° , i = 10°, b = 3 mm, t = 0.25 mm, V_w = 30 m/min, k = 350 N/mm^2, μ = 0.577 and ϕ_n = 25°. Assuming Stabler chip-flow rule, calculate the cutting-power requirement.

Solution :

Cutting force,

$$F_p = \frac{kbt}{\sin\phi_n}\left\{\frac{\cos(\lambda_n - \alpha_n) + \tan i \tan\eta_c \sin\lambda_n}{\sqrt{\cos^2(\phi_n + \lambda_n - \alpha_n) + \tan^2\eta_c \sin^2\lambda_n}}\right\}$$

$$= \frac{350\times3\times0.25}{\sin25^\circ}\left\{\frac{\cos(30^\circ-20^\circ)+\tan10^\circ\tan10^\circ\sin30^\circ}{\sqrt{\cos^2(25^\circ+30^\circ-20^\circ)+\tan^2 10^\circ\sin^2 30^\circ}}\right\}$$

$$= \frac{350\times3\times0.25}{0.42}\left\{\frac{0.98+0.176\times0.176\times0.5}{\sqrt{0.67+0.03\times0.25}}\right\} = 755.895 \text{ N}$$

since

$$\tan\lambda_n = 0.577$$

or

$$\lambda_n = 30^\circ$$

and

$$\eta_c = i = 10^\circ \text{ (Stabler chip-flow rule)}$$

Cutting power,

$$P_W = F_P \times V$$
$$= 755.895 \times 30 \text{ Nm/min}$$
$$= 22668.75 \text{ Nm/min}$$

Example 15. A two-component cantilever beam dynamometer has to be used for measuring the forces. The cross-section of the beam is 10 mm × 10 mm and the strain gauges are 75 mm away from the cutting edge. Calculate the millivolt output from the bridges per kg when 120 ohm strain gauges having gauge factor of 2 are used and the bridges are fed with a 12 volts dc supply. Assume all arms of the bridges to be active ($E = 20 \times 10^4$N/mm).

Solution:

Equations (6.4) and (6.6) give

$$e_1 = e_2 = e_3 = e_4 = \frac{6F_q l}{Ebh^2}$$

and

$$e_1' = e_2' = e_3' = e_4' = \frac{6F_p l}{Ebh^2}$$

For 1 kg f = 10 N force,

$$e = \frac{6\times10\times75}{20\times10^4\times10\times10\times10} = 22.5\times10^{-6}$$

Using equation (6.23), the output voltage can be evaluated as

$$v' = \frac{n}{4}GV_e e$$

$$= \frac{4}{4}\times2\times12\times22.5\times10^{-6}$$

$$= 0.54 \text{ millivolts}$$

Example 16. An extended ring of radius 30 mm, width 75 mm and thickness 2 mm is to be used as a milling dynamometer. What is the maximum force it can measure without damaging the strain gauges ? (Take $E = 21 \times 10^4$ N/mm^2)

Solution:

Wire resistance gauges are seldom subjected to strain exceeding 0.001 in dynamometer application. Using this value in equations (6.8) and (6.9), we get

$$e_{90^\circ} = \pm 1.09 \frac{F_q r}{Ebt^2}$$

or

$$(F_q)_{\max} = \frac{0.001 \times 21 \times 10^4 \times 75 \times 2 \times 2}{1.09 \times 30} = 1926.6 \text{ N}$$

and

$$e_{39.6^\circ} = \pm 2.31 \frac{F_p r}{Ebt^2}$$

or

$$(F_p)_{\max} = \frac{0.001 \times 21 \times 10^4 \times 75 \times 2 \times 2}{2.31 \times 30} = 909 \text{ N}$$

Example 17. Evaluate the thickness of the octagonal ring of an extended octagonal ring grinding dynamometer to provide a bridge output voltage of 0.1 millivolt at $\pm$1N force in the vertical direction when the input voltage is 6 volts. The gauge factor of the gauges used is 2.8 and E for the material is 22×10^4 N/mm^2. The radius of the ring should not exceed 50 mm. The width may be taken as 30 mm.

Solution:

From equation (6.13),

$$e_{90^\circ} = \pm 0.7 \frac{F_q r}{Ebt^2}$$

and equation (6.23) gives

$$v' = \frac{n}{4} G V_e \, e$$

or

$$e = \frac{4v'}{nGV_e} = \frac{4 \times 0.1 \times 10^{-3}}{4 \times 2.8 \times 6}$$

$$= 5.95 \times 10^{-6}$$

Now

$$t^2 = \frac{0.7 F_q r}{Ebe}$$

$$= \frac{0.7 \times 1 \times 50}{22 \times 10^4 \times 30 \times 5.95 \times 10^{-6}} = 0.8913$$

Therefore,

$$t = 0.944 \text{ mm}$$

Example 18. Estimate the average shear plane temperature for machining mild steel at a cutting speed of 2 m/sec with a 10° rake angle tool, the width of cut being 2 mm and the undeformed chip-thickness 0.25 mm. Given $\rho = 7200$ kg/m^3, $c = 502$ J/kg °C, $\mu = 0.5$ and $k = 400 \times 10^6$ N/m^2.

Solution:

Merchant's shear angle relationship (equation 4.9) gives

$$\phi = \frac{\pi}{4} - \frac{1}{2}(\lambda - \alpha)$$

or

$$\phi = \frac{\pi}{4} - \frac{1}{2}(\tan^{-1} 0.5 - 10°) = 36.72°$$

From equations (4.2) and (4.3)

$$F_s = \frac{kbt}{\sin\phi}$$

$$= \frac{400 \times 10^6 \times 2 \times 10^{-3} \times 0.25 \times 10^{-3}}{\sin 36.72°}$$

$$= 334.6 \text{ N}$$

From equation (4.21),

$$V_s = \frac{\cos\alpha}{\cos(\phi - \alpha)} \cdot V$$

$$= \frac{\cos 10°}{\cos 26.72°} \cdot 2 = 2.21 \text{ m/sec}$$

Equation (7.27) can now be used for estimating the average shear plane temperature. Thus,

$$\bar{\theta}_s = \frac{\beta F_s V_s}{\rho t b c V} + \theta_o$$

$$= \frac{0.85 \times 334.6 \times 2.21}{7200 \times 0.25 \times 10^{-3} \times 2 \times 10^{-3} \times 502 \times 2} + 25$$

$$= 173.9 + 25 = 198.9° \text{ C}$$

Here β has been assumed to be 0.85.

Example 19. Evaluate the average shear plane and rake face temperatures during orthogonal cutting with a zero rake tool when $V = 2$ m/sec, $b = 2$ mm and $t = 0.2$ mm. The chip-thickness ratio and chip-tool contact length obtained at these conditions were 0.2 and 0.45 mm, respectively. The dynamometer measurement gave $F_p = 760$ N and $F_q = 390$ N. ($\rho = 7200$ kg/m^3, $c = 502$ J/kg °C, $K = 50$ W/m °C, $\theta_o = 25$ °C, $\beta = 0.85$).

Solution:

From equation (7.27)

$$\bar{\theta}_s = \frac{\beta F_s V_s}{\rho t b c V} + \theta_o$$

Equation (4.21) gives

$$V_s = \frac{\cos\alpha}{\cos(\phi-\alpha)} V$$

and equation (4.28) gives

$$\tan\phi = \frac{r\cos\alpha}{1-r\sin\alpha} = \frac{0.2\cos 0^\circ}{1-0.2\sin 0^\circ} = 0.2$$

or

$$\phi = \tan^{-1} 0.2 = 11.31^\circ$$

Now,

$$F_s = F_p\cos\phi - F_q\sin\phi$$

$$= 760 \cos 11.31^\circ - 390 \sin 11.31^\circ = 668.76 \text{ N}$$

Therefore,

$$\bar{\theta}_s = \frac{0.85\times 668.76\times(\cos 0^\circ/\cos 11.31^\circ)\times V}{7200\times 0.2\times 10^{-3}\times 2\times 10^{-3}\times 502\times V} + 25$$

$$= 401 + 25 = 426^\circ \text{ C}$$

From equation (7.34)

$$\bar{\theta}_f = \frac{0.678F\sqrt{V_c}\cdot\sqrt{\alpha_d}}{\sqrt{at_c}\cdot K b} + \bar{\theta}_s$$

Equation (4.21) gives

$$V_c = \frac{\sin\phi}{\cos(\phi-\alpha)} V = \frac{\sin 11.31^\circ}{\cos 11.31^\circ}\times 2$$

$$= 0.4 \text{ ms}^{-1}$$

Using equation (4.12), we get the friction force

$$F = F_p\sin\alpha + F_q\cos\alpha = 390 \text{ N}$$

Thermal diffusivity,

$$\alpha_d = \frac{K}{\rho c} = \frac{50}{7200\times 502} = 1.38\times 10^{-5} \text{ m}^2/\text{s}$$

The chip-tool contact length,

$$at_c = 0.45 \text{ mm}$$

Substituting these values, we get

$$\bar{\theta}_f = \frac{0.678\times 390\times\sqrt{0.4}\sqrt{1.38\times 10^{-5}}}{\sqrt{0.45\times 10^{-3}}\times 50\times 2\times 10^{-3}} + 426$$

$$= 292.86 + 426 = 718.86 \text{ °C}.$$

Example 20. During straight turning of a 25 mm diameter steel bar at 300 rpm with a HSS tool, a tool life of 10 min was obtained. When the same bar was turned at 250 rpm, the tool life increased to 52.5 min. What will be the tool life at a speed of 275 rpm?

Solution:

In this problem,

$$V_1 = \frac{\pi \times 25 \times 300}{1000} \text{ m/min; } T_1 = 10 \text{ min}$$

$$V_2 = \frac{\pi \times 25 \times 250}{1000} \text{ m/min; } T_2 = 52.5 \text{ min}$$

Using Taylor's tool life equation,

$$V_1 T_1^n = V_2 T_2^n$$

we get

$$n = \frac{\log(V_2/V_1)}{\log(T_1/T_2)} = \frac{\log\frac{250}{300}}{\log\frac{10}{52.5}} = 0.11$$

Now,

$$V_1 T_1^n = V_3 T_3^n$$

or

$$T_3 = \left(\frac{V_1 T_1^n}{V_3}\right)^{1/n} = \left(\frac{V_1}{V_3}\right)^{1/n} T_1$$

$$= \left(\frac{300}{275}\right)^{1/0.11} \times 10 = 22 \text{ min}$$

Example 21. During machining with a carbide tool, the wear land size l_w (mm) was found to be related to the cutting speed V (m/min) and tool life T (min) by the following relationship:

$$l_w = 56.5 \times 10^{-8} V^{2.4} T^{0.6}$$

Assuming $l_w^* = 1.5$ mm as the criterion for tool failure, obtain the tool-life equation and evaluate the cutting speed for a tool life of 30 min.

Solution:

When $l_w = l_w^* = 1.5$ mm, we get

$$V^{2.4} T^{0.6} = \frac{1.5}{56.5 \times 10^{-8}}$$

or

$$VT^{0.6/2.4} = \left(\frac{1.5}{56.5 \times 10^{-8}}\right)^{1/2.4}$$

or

$$VT^{0.25} = 475$$

This is the tool-life equation.

For $T = 30$ min,

$$[V]_{30\,\text{min}} = \frac{475}{30^{0.25}} = 202.96 \text{ m/min}$$

Example 22. During machining of a steel, tool A and tool *B* gave $VT^{0.12} = 60$ and $VT^{0.22} = 85$ as tool-life equations. Evaluate the break-even-speed above which tool *B* will give better tool-life.

Solution:

At break-even-speed V^*, both tools will give the same tool life, Therefore,

$$V^* = \frac{60}{T^{0.12}} = \frac{85}{T^{0.22}}$$

or

$$T^{0.1} = \frac{85}{60}$$

or

$$T = 32.6 \text{ min}$$

The break-even-speed can now be obtained as

$$V^* = \frac{60}{32.6^{0.12}} = 39.5 \text{ m/min}$$

At speeds above this, tool *B* will give better tool life.

Example 23. During turning of mild steel at a cutting speed of 40 m/min with an HSS tool of 0-0-6-6-8-30-1 mm (ASA) the tool life was obtained as 40 minutes. What will be the tool life at 60 m/min for an HSS tool of 0-0-6-6-8-15-1 mm (ASA)? The constant *C* of the Taylor's tool life equation is related to the side cutting-edge angle α_s (degrees) as

$$C = 78\ (\alpha_s + 15)^{0.26}$$

Solution:

For side cutting-edge angle of 30°,

$$C = 78\ (30 + 15)^{0.26} = 209.86$$

and for side cutting edge angle of 15°,

$$C = 78\ (15 + 15)^{0.26} = 188.86$$

Using Taylor's tool life eqn. $VT^n = C$ (equation 8.1), the tool life exponent *n* may be expressed as

$$n = \log\left(\frac{C}{V}\right)/\log T = \log\left(\frac{209.86}{40}\right)/\log(40) = 0.449$$

Thus tool life,

$$T = \left(\frac{188.86}{60}\right)^{\frac{1}{0.449}} = 12.86 \text{ min}$$

Example 24. A wedge-shaped tool with rake angle $\alpha = 0°$ and flank angle $\delta = 2°$ gave a tool life of T min. Assuming the tool wear rate normal to the wear land to be constant, evaluate the increase in tool life when the flank angle is increased to 8°.

Solution:

Since the rake angle is 0° and the wear rate normal to the wear land is constant,

$$T \propto l_w^* \tan\delta$$

where l_w^* is the wear-land width at tool failure. This gives,

$$\frac{(T)_{\delta=8°}}{(T)_{\delta=2°}} = \frac{\tan 8°}{\tan 2°} = 4$$

The tool life at $\delta = 8°$ will be 4 times the tool life at $\delta = 2°$ for a given wear-land criterion.

Example 25. In a facing tool life test, the wear limit (critical wear land l_w^*) of 1 mm was reached at 200 mm diameter at 250 rpm when the feed rate was 0.2 mm/rev. If the Taylor tool life exponent n is 0.20, what is the value of constant C ? (Assume the cutting to start from the centre of the plate)

Solution:

Equivalent speed,

$$V_{eq} = \pi N\left[\frac{n}{1+n}\left(\frac{D_2^{\frac{1+n}{n}} - D_1^{\frac{1+n}{n}}}{D_2 - D_1}\right)\right]^n$$

$$= \frac{\pi \times 250}{1000}\left[\frac{0.2}{1.2}\left(\frac{200^6}{200}\right)\right]^{0.2} = 109.6 \text{ m/min}$$

since $D_1 = 0$.

$$T = \frac{D_2 - D_1}{2fN}$$

$$= \frac{200}{2 \times 0.2 \times 250} = 2 \text{ min}$$

Now,

$$V_{eq}T^{0.2} = C$$

or

$$C = (109.6) \times (2)^{0.2} = 125.9$$

Example 26. In a taper-turning tool-life test at speeds N_1 and N_2 rpm, the fraction of tool life consumed X_1 and X_2 were recorded as 0.48 and 0.35, respectively. The corresponding cutting times T_1 and T_2 were 1.75 and 2.3 min. Taking the length of cut L to be 300 mm and feed f as 0.2 mm/rev, calculate the Taylor exponent n.

Solution:

For taper turning,

$$T_1 = \frac{L}{fN_1}$$

or

$$N_1 = \frac{L}{fT_1} = \frac{300}{0.2 \times 1.75} = 857.1 \text{ rpm}$$

and

$$N_2 = \frac{L}{fT_2} = \frac{300}{0.2 \times 2.3} = 652.2 \text{ rpm}$$

Tool-life exponent,

$$n = \frac{1}{\left(\log\frac{X_1}{X_2} / \log\frac{N_1}{N_2}\right) + 1}$$

$$= \frac{1}{\left(\log\frac{0.48}{0.35} / \log\frac{857.1}{652.2}\right) + 1} = \frac{1}{2.15} = 0.46$$

Example 27. A plain milling cutter 80 mm in diameter with 8 teeth is rotating at 60 rpm. If the feed rate provided is 50 mm/min, estimate the peak-to-valley roughness.

Solution:

Equation (10.40) gives the peak-to-valley roughness in plain milling as

$$h = \frac{v^2}{4Dn^2N^2} = \frac{50 \times 50}{4 \times 80 \times 8 \times 8 \times 60 \times 60}$$

$$= 33.91 \times 10^{-6} \text{ mm}$$

Example 28. Estimate the longitudinal and vertical force components F_p and F_q during plain milling of a 30 mm wide mild steel block with a straight teeth cutter of 40 mm diameter having 10 teeth. The rpm of the cutter is 60, the feed rate is 30 mm/min and the depth of cut is 1 mm. The rake face coefficient of friction may be taken as 0.5 and the yield shear stress of work material as 400 N/mm^2. The radial rake angle of the cutter is 10°.

Solution:

The average undeformed chip-thickness in this case is given by equation (10.30). Thus,

$$\bar{t} = \frac{v}{nN}\sqrt{(d/D)}$$

$$= \frac{30}{10 \times 60}\sqrt{\frac{1}{40}} = 0.0079 \text{ mm}$$

Using Merchant's theory (equation 4.9) gives,

$$\phi = \frac{\pi}{4} - \frac{1}{2}(\lambda - \alpha)$$

Here

$$\alpha = \alpha_r = 10°$$

and

$$\lambda = \tan^{-1} 0.5 = 26.6°$$

Therefore,

$$\phi = \frac{\pi}{4} - \frac{1}{2}(26.6° - 10°) = 36.7°$$

We can now use equations (4.6) and (4.7) to estimate the force components F_p and F_q. Therefore,

$$F_p = \frac{kbt\cos(\lambda - \alpha)}{\sin\phi\cos(\phi + \lambda - \alpha)}$$

$$= \frac{400 \times 30 \times 0.0079 \times \cos \times 16.6°}{\sin 36.7° \times \cos 53.3°} = 254.37 \text{ N}$$

and

$$F_q = \frac{kbt\sin(\lambda - \alpha)}{\sin\phi\cos(\phi + \lambda - \alpha)}$$

$$= \frac{400 \times 30 \times 0.079 \times \sin 16.6°}{\sin 36.7° \times \cos 53.3°} = 75.83 \text{ N}$$

Example 29. A cemented carbide tool is turning an alloy steel rod of diameter 50 mm and length 150 mm at a depth of cut of 3 mm and a feed of 0.4 mm/rev. The tool life equation is $VT^{0.2} = 200$, where V is in m/min and T is in min. The tool changing time is 1 min and the time for loading and unloading of workpiece as well as advance and withdrawal of tool is 1 min. Initial set-up and idle time need not be considered. The tool regrinding cost is 3.0 units of cost and a new tool costs 100 units of cost. Tools can be reground a total of 10 times. Labour, overhead and depreciation is 15 units of cost per hour. What is the minimum cost per piece and the maximum production rate? What will be *(i)* the cost per piece at the maximum production rate and *(ii)* production rate at minimum cost?

Solution:

Substituting for T_c and N_g in equation (9.8) and equating $\frac{dC_B}{dV} = 0$ gives equation (9.26) for optimum cutting speed for minimum cost as

$$V_1^* = \left[\frac{nK_1K_2(c_L + c_O + c_D)}{(1-n)(c_L + c_O + c_D)T_r + \frac{C_i}{r_g} + C_g}\right]^n$$

Now

$$n = 0.2$$

$$K_1 = \frac{\pi DL}{f} = \frac{\pi \times 0.05 \times 0.15}{0.4 \times 10^{-3}} = 58.9$$

$$K_2 = \frac{C^{1/n}}{K_1} = \frac{200^{1/0.2}}{58.9} = 5.4 \times 10^9$$

$$(c_L + c_O + c_D) = 15 \text{ units of cost/hour}$$
$$= 0.25 \text{ units of cost/min}$$
$$T_r = 1 \text{ min.}$$
$$C_i = 100 \text{ units of cost}$$
$$C_g = 3 \text{ units of cost}$$
$$r_g = 10$$

Therefore,

$$V_1^* = \left[\frac{0.2 \times 58.9 \times 5.4 \times 10^9 \times 0.25}{(1-0.2) \times 0.25 \times 1 + \dfrac{100}{10} + 3}\right]^{0.2} = 65.5 \text{ m/min}$$

The minimum cost per piece can now be evaluated using equations (9.22) as

$$(C_p)_{\min} = \frac{1}{N_B}\left\{(c_L + c_O + c_D)\{N_B(T_1 + T_a) + T_s + T_i\} + \frac{N_B (V_1^*)^{\frac{1-n}{n}}}{K_2}\right.$$
$$\left. \times \left\{(c_L + c_O + c_D)T_r + \frac{C_i}{r_g} + C_g\right\} + \left\{(c_L + c_O + c_D) N_B K_1 (V_1^*)^{-1}\right\}\right.$$

$$= \frac{1}{N_B}[0.25 \times \{N_B \times 1 + 0 + 0\} + N_B \frac{(65.5)^4}{5.4 \times 10^9}\{0.25 \times 1 + \frac{100}{10} + 3\}$$

$$+ \{0.25 \times N_B \times 58.9 \times (65.5)^{-1}\}]$$
$$= [0.25 + 0.045 + 0.225] = 0.52 \text{ units of cost}$$

The optimum cutting speed for maximum production rate or minimum production time V_2^* can be obtained from cquation (9.29) as

$$V_2^* = \left[\frac{n K_1 K_2}{(1-n)T_r}\right]^n$$

$$= \left[\frac{0.2 \times 58.9 \times 5.4 \times 10^9}{0.8 \times 1}\right]^{0.2} = 151.4 \text{ m/min.}$$

The maximum production rate is then obtained by substituting V_2^* in place of V in equation (9.27). Thus,

$$(R_p)_{\max} = \frac{1}{\left\{\frac{1}{N_B}(T_s + T_i)\right\} + \left\{T_1 + T_a + K_1(V_2^*)^{-1} + \frac{T_r(V_2^*)^{\frac{1-n}{n}}}{K_2}\right\}}$$

$$= \frac{1}{\{0\} + \left\{(1 + 58.9 \times (151.4)^{-1} + \frac{1 \times (151.4)^4}{5.4 \times 10^9}\right\}}$$

$$= 0.67 \text{ pieces/ min or } 40.3 \text{ pieces/hour}$$

Cost per piece $(C_p)_{MPR}$ at maximum production rate is

$$(C_p)_{MPR} = \frac{1}{N_B}\left[(c_L + c_O + c_D)\{N_B(T_1 + T_a) + T_s + T_i\} + \frac{N_B(V_2^*)^{\frac{1-n}{n}}}{K_2}\right.$$

$$\left. \times \left\{(c_L + c_O + c_D)T_r + \frac{C_i}{r_g} + C_g\right\} + \left\{(c_L + c_O + c_D)N_B K_1(V_2^*)^{-1}\right\}\right]$$

$$= \frac{1}{N_B}\left[0.25 \times \{N_B \times 1 + 0 + 0\} + N_B \frac{(151.4)^4}{5.4 \times 10^9}\left\{0.25 \times 1 + \frac{100}{10} + 3\right\}\right.$$

$$\left. + \{0.25 \times N_B \times 58.9 \times (151.4)^{-1}\}\right]$$

$$= (0.25 + 1.29 + 0.097) = 1.63 \text{ units of cost}$$

Production rate $(R_p)_{MC}$ at minimum cost per piece is

$$(R_p)_{MC} = \frac{1}{\left\{\frac{1}{N_B}(T_s + T_i)\right\} + \left\{T_1 + T_a + K_1(V_2^*)^{-1} + \frac{T_r(V_1^*)^{\frac{1-n}{n}}}{K_2}\right\}}$$

$$= \frac{1}{\{0\} + \left\{(1 + 58.9 \times (65.5)^{-1} + \frac{1 \times (65.5)^4}{5.4 \times 10^9}\right\}}$$

$$= 0.53 \text{ pieces/min. or } 31.6 \text{ pieces/hour}$$

Example 30. Evaluate the tool life for maximum production rate when the Taylor tool life exponent n is 0.2 and the tool changing time is 1.5 min.

Solution:

Equation (9.29) gives the optimum cutting speed for maximum production rate as

$$V_2^* = \left[\frac{nK_1K_2}{(1-n)T_r}\right]^n$$

or

$$\frac{C}{T^n} = \left[\frac{nK_1C^{1/n}/K_1}{(1-n)T_r}\right]^n$$

or

$$T = \left(\frac{1-n}{n}\right)T_r$$

where T is the tool life at V_2^*.

Substituting for n and T, we get

$$T = \left(\frac{1-0.2}{0.2}\right)\times 1.5 = 6 \text{ min}$$

Example 31. During machining of a steel, tool I gave $VT^{0.1} = 60$ while tool II gave $VT^{0.16} = 85$ as tool life equations. The tool changing time for both tools was 3 min. Which tool will give a higher production rate?

Solution:

The optimum cutting speed for maximum production rate (equation 9.29) is

$$V_1^* = \left[\frac{nK_1K_2}{(1-n)T_r}\right]^n$$

or

$$V_1^* = \left[\frac{n}{(1-n)T_r}\right]^n C$$

Now

$$[V_2^*]_{\text{Tool I}} = \left[\frac{0.1}{(1-0.1)\times 3}\right]^{0.1}\times 60$$

$$= 43.2 \text{ m/min.}$$

and

$$[V_2^*]_{\text{Tool II}} = \left[\frac{0.16}{(1-0.16)\times 3}\right]^{0.16}\times 85$$

$$= 54.7 \text{ m/min}$$

Since tool II gives a higher value of optimum speed, higher production rate will be obtained with this tool.

Example 32. During plunge-cut surface grinding of a 10 mm wide workpiece with a 200 mm × 20 mm wheel at a table speed of 2.5 m/min and wheel depth of cut of 0.02 mm the tangential and thrust forces were recorded as 20 N and 38 N. The grain density was separately measured to be 250 per cm^2 and the wheel rpm was 3000. Evaluate the specific energy and the chip-thickness (Take r = 15).

Solution:

Given

$$D = 200 \text{ mm}$$
$$B = 10 \text{ mm}$$
$$v = 2.5 \text{ m/min}$$
$$d = 0.02 \text{ mm}$$
$$C = 250/\text{cm}^2$$
$$F_p = 20 \text{ N}$$
$$N = 3000 \text{ rpm}$$

From equation (11.20), the specific energy u can be obtained. Thus,

$$u = \frac{F_p V}{vBd}$$

$$= \frac{20 \times \pi \times 200 \times 3000}{2.5 \times 10^3 \times 10 \times 0.02}$$

$$= 7.54 \times 10^4 \text{ N/mm}^2$$

Chip thickness t can be evaluated from equation (11.18) as

$$t = \left[\frac{4v}{VC_r}\sqrt{\frac{d}{D}}\right]^{1/2}$$

$$= \left[\frac{4 \times 2.5 \times 10^3}{\pi \times 200 \times 3000 \times 250 \times 10^{-2} \times 15} \frac{\sqrt{0.02}}{200}\right]^{1/2}$$

$$= 1.19 \times 10^{-3} \text{ mm}$$

Example 33. The following data were recorded during plunge-cut grinding of 10 mm wide workpiece with a 300 mm × 20 mm wheel rotating at 3000 rpm:

	Work speed (m/min)	Wheel depth of cut (mm)	Tangential force (N)	Thrust force (N)
1.	2	0.02	20	38
2.	4	0.03	40	81

The grain density C and r are taken to be 300 per cm^2 and 15. Evaluate the exponent m relating the specific energy u with chip thickness t.

Solution:

Using equation (11.18), chip thickness t can be evaluated as

$$t = \left[\frac{4v}{VCr}\sqrt{\frac{d}{D}}\right]^{1/2}$$

Thus,

$$t_1 = \left[\frac{4\times 2}{\pi\times 300\times 10^{-3}\times 3000\times 300\times 10^{-2}\times 15}\sqrt{\frac{0.02}{300}}\right]^{1/2}$$

$$= 0.00071 \text{ mm}$$

and

$$t_2 = \left[\frac{4\times 4}{\pi\times 300\times 10^{-3}\times 3000\times 300\times 10^{-2}\times 15}\sqrt{\frac{0.03}{300}}\right]^{1/2}$$

$$= 0.00112 \text{ mm}$$

The specific energy can be evaluated from equation (11.20) as

$$u = \frac{F_p V}{vBd}$$

Thus,

$$u_1 = \frac{20\times\pi\times 300\times 10^{-3}\times 3000}{2\times 10\times 0.02}$$

$$= 1.414 \times 10^5 \text{ N/mm}^2$$

and

$$u_2 = \frac{40\times\pi\times 300\times 10^{-3}\times 3000}{4\times 10\times 0.03}$$

$$= 9.425 \times 10^4 \text{ N/mm}^2$$

The specific energy in grinding is related to chip thickness (equation 11.21) as $u \propto \frac{l}{t^m}$. Therefore,

$$u_1\, t_1^{\,m} = u_2\, t_2^{\,m}$$

or

$$m = \frac{\log(u_1/u_2)}{\log(t_2/t_1)}$$

Substituting, we get

$$m = \frac{\log(1.414\times 10^5/9.425\times 10^4)}{\log(0.00112/0.00071)} = 0.89$$

Example 34. Estimate the mean undeformed chip-thickness and the ideal surface roughness during plain milling under the following conditions:

Cutter diameter	:	75 mm
No. of teeth	:	8
Cutter speed	:	60 rpm
Feed	:	100 mm/min
Depth of cut	:	2 mm

Solution:

The mean undeformed chip-thickness in plain milling can be estimated from equation (10.30) as

$$t = \frac{v}{nN}\sqrt{\left(\frac{d}{D}\right)}$$

$$= \frac{100\times10^{-3}}{8\times60}\times\left(\frac{2\times10^{-3}}{75\times10^{-3}}\right)^{1/2} = 34 \times 10^{-6} \text{ mm}$$

The ideal surface roughness can be estimated from equation (10.40). Thus,

$$h = \frac{v^2}{4Dn^2N^2}$$

$$= \frac{1}{4\times75\times10^{-3}}\times\left[\frac{100\times10^{-3}}{8\times60}\right]^2$$

$$= 0.145 \times 10^{-6} \text{ mm}$$

Example 35. Evaluate the peak-to-valley roughness of a flat surface ground under the following conditions:

Grinding wheel size	:	300 mm × 30 mm
No. of cutting points/mm^2	:	3
Wheel rpm	:	2500
Work speed	:	100 mm/min
Depth of cut	:	0.05 mm

The value of the ratio of mean chip-width to mean chip-thickness may be taken as 15.

Solution:

The mean peak-to-valley roughness can be evaluated from equation (11.33) as

$$h = \frac{t^2}{16d}$$

where

$$t = \left[\frac{4v}{VCr}\sqrt{(d/D)}\right]^{1/2}$$

from equation (11.18). Thus,

$$t = \left[\frac{4\times(100\times10^{-3}/60)}{(\pi\times300\times10^{-3}\times2500/60)\times3\times15}\times\sqrt{\left(\frac{0.05}{300}\right)}\right]^{1/2}$$

$$= 2.2 \times 10^{-4} \text{ mm}$$

and

$$h = \frac{(2.2\times10^{-4})^2}{16\times0.05} = 0.061 \times 10^{-6} \text{ mm}$$

Example 36. Estimate the average workpiece temperature during horizontal surface grinding of a 10 mm wide workpiece with a 250 mm × 20 mm grinding wheel rotating at 3000 rpm. The following data are provided:

Wheel depth of cut	:	0.05 mm
Table speed	:	100 mm/min
Tangential grinding force	:	6 N
Thermal conductivity of workpiece material	:	50 W/m°C
Thermal diffusivity of workpiece material	:	1.6×10^{-5} m²/sec

Solution:

The average workpiece surface temperature can be evaluated using equation (11.27) as

$$\theta = \frac{0.754 q l'}{K\sqrt{L'}}$$

where

$$q = \bar{q} = 0.8 \frac{F_P V}{B l_s}$$

$$l' = \frac{l_s}{2}$$

and

$$L' = \frac{l_s v}{2\alpha_d}$$

Now

$$l_s = \sqrt{(dD)} = \sqrt{(0.05 \times 250)/1000}$$

$$= 3.54 \times 10^{-3} \text{ m}$$

Hence,

$$\bar{q} = \frac{0.8 \times 6 \times (\pi \times 250 \times 3000 \times 10^{-3}/60)}{10 \times 10^{-3} \times 3.54 \times 10^{-3}}$$

$$= 5.32 \times 10^6 \text{ N m/sec m}^2$$

$$l' = \frac{3.54 \times 10^{-3}}{2} = 1.77 \times 10^{-3} \text{ m}$$

$$L' = \frac{3.54 \times 10^{-3} \times (100 \times 10^{-3}/60)}{2 \times 1.6 \times 10^{-5}} = 0.184$$

Therefore,

$$\bar{\theta} = \frac{0.754 \times 5.32 \times 10^6 \times 1.77 \times 10^{-3}}{50 \times \sqrt{0.184}} = 331\ ^\circ\text{C}$$

Review Questions

1 Explain how chips are formed in metal cutting?

2 Differentiate between orthogonal and oblique cutting.

3 Explain how material transfer takes place during built-up-edge formation?

4 Under what machining conditions, a ductile material like aluminium will give discontinuous chips?

5 Distinguish between generation and tracing methods of machining.

6 Show that $\tan \gamma_s = \tan \alpha_b \cot \alpha_s$ for orthogonal cutting. Here γ_s is the side cutting-edge angle and α_b and α_s are the back and side rake angles.

7 Prove that $\tan i = \tan \alpha_b \cos \gamma_s - \tan \alpha_s \sin \gamma_s$, where i is the inclination angle, γ_s is the side cutting-edge angle and α_b and α_s are the back and side rake angles.

8 Why cutting-edge angles are provided on single-point tools?

9 Sketch the Merchant's circle diagram and state the assumptions made in developing this diagram.

10 Discuss the friction process on the rake face.

11 The shear stress value of work material evaluated from experimental metal cutting data is usually much higher than that obtained by conventional materials test. Why?

12 How shear angle can be evaluated from chip-length measurements?

13 Prove that

$$F_p V = F_s V_s + F V_c$$

during orthogonal cutting, here F, F_p and F_s are friction, tangential and shear force components and V, V_c and V_s are work, chip and shear velocities.

14 Under what conditions Lee and Shaffer's shear angle relationship $\phi = \frac{\pi}{4} - (\lambda - \alpha)$ is not valid? Explain why?

15 Show that during orthogonal machining with a zero rake angle tool the shear strain $\gamma = \frac{1+r^2}{r}$, where, r is the chip-thickness ratio.

16 Show that the specific energy u_s during orthogonal cutting with a zero rake angle tool is given by

$$u_s = \frac{k(1+r^2)}{r(1-\mu r)}$$

where k is the yield shear stress of work material, μ is the co-efficient of friction and r is the chip-thickness ratio.

17 Show that in orthogonal cutting, the shear strain γ can be evaluated from

$$\gamma = \frac{1-2r\sin\alpha + r^2}{r\cos\alpha}$$

where α is the rake angle and r is the chip-thickness ratio.

18 Describe the method adopted by Oxley for obtaining the shear angle in orthogonal cutting.

19 What basic assumptions were made by Oxley for developing his thin-zone orthogonal cutting model and how did he verify them?

20 Sketch an oblique cutting set-up and indicate how α_v, α_e and α_n are defined? Which one of these angles is generally considered to be important and why?

21 What is Stabler's chip-flow rule? List two other chip-flow rules that have been suggested.

22 What are parasitic forces? Describe Spaans method of determining these forces.

23 Discuss the requirements of machine-tool dynamometers.

24 Why octagonal rings are preferred over circular rings for machine-tool dynamometers although circular rings are more effective in minimizing cross-sensitivity?

25 What are the heat sources in metal cutting and what are their effects?

26 How chip-tool contact temperature can be measured?

27 Show that during orthogonal cutting with a tool of zero rake angle the rate of theat generation on the shear plane is equal to $P_w\ (1-\mu r)$, where P_w is the cutting power, μ is the co-efficient of friction and r is the chip-thickness ratio.

28 Discuss the basic causes of tool wear.

29 Explain how the adhesion phenomenon has been used to qualitatively explain the wear-land growth curve.

30 Discuss the tool life and temperature relationship.

31 Cutting tool material is much harder than the workpiece, yet the tool wears during machining. Explain why?

32 Why it is important to indicate the tool failure criterion while specifying tool life?

33 Why crater is formed some distance above the tool tip on the rake face?

34 Explain the most common tool failure criteria.

35 Discuss the important factors that affect tool life.

36 How Taylor's tool-life equation can be obtained experimentally?

37 List the common single-point cutting tool materials and indicate how they meet the basic requirements.

38 How cutting fluids are effective in improving tool life?

39 How effects of variable speed are interpreted in terms of equivalent speed in quick tool-life-testing?

40 How Taylor tool life exponent n is obtained from taper turning tests?

41 Discuss machinability as a parameter to describe the machining properties of work materials.

42 Discuss the various criteria used for optimizing the cutting conditions.

43 Why is the optimum cutting speed for minimum cost always less than that for maximum production rate?

44 Show that in a single-pass straigh-turning operation the tool life for maximum production rate is equal to $T_r (1 - n)/n$, where n is the tool life exponent and T_r is the tool removing and replacing time.

45 What are the advantages in using helical milling cutters?

46 Discuss the process of chip formation with a twist drill.

47 What cutting variables affect the surface finish in plain milling operations? How should these variables be changed to obtain better finish?

48 The ideal surface roughness values obtained theoretically are generally lower than the values obtained experimentally. Why?

49 How are grinding wheels specified?

50 How the performance of a grinding is generally evaluated?

51 Two wheels of the same specifications behave differently when they are of different diameters. Why?

52 Sketch the wheel-wear pattern in fine and rough grinding and discuss the behaviour.

53 How the number of cutting points on the wheel surface is measured?

54 Why specific energy in grinding is much higher than in single-point cutting?

55 Discuss the wear mechanism of a grinding wheel.

56 What is wheel loading and glazing? What considerations are made at the time of wheel selection to minimize these?

57 How should cutting conditions be changed to improve the surface finish in surface grinding without changing the wheel?

58 What are different types of machining vibrations and their causes?

59 Explain how machining vibrations occur due to regeneration effect.

60 Describe how mode-coupling effect causes machining vibration.

Conversion to SI Units

1. Length	:	inch	= metre	$\times (2.54 \times 10^{-2})$
	:	foot	= metre	$\times (3.048 \times 10^{-1})$
2. Mass	:	pound	= kilogram	$\times (4.536 \times 10^{-1})$
3. Time	:	minute	= second	$\times (6.0 \times 10^{-1})$
	:	hour	= second	$\times (3.6 \times 10^{3})$
4. Area	:	square inch	= square metre	$\times (6.452 \times 10^{-4})$
5. Volume	:	cubic inch	= cubic metre	$\times (1.639 \times 10^{-5})$
6. Force	:	pound	= newton	$\times (4.448)$
7. Pressure/Stress	:	pounds per square inch	= newton per square metre (or pascal)	$\times (6.894 \times 10^{3})$
8. Density	:	pound per cubic inch	= kilogram per cubic metre	$\times (2.768 \times 10^{4})$
9. Speed	:	foot per minute	= metre per second	$\times (5.08 \times 10^{-3})$
	:	inch per minute	= metre per second	$\times (4.233 \times 10^{-4})$
	:	revolution per minute	= radians per second	$\times (1.047 \times 10^{-1})$
10. Acceleration	:	foot per second per second	= metre per second per second	$\times (3.048 \times 10^{-1})$
	:	inch per second per second	= metre per second per second	$\times (2.54 \times 10^{-2})$
11. Work/Energy	:	BTU	= netwon-metre (or joule)	$\times (1.055 \times 10^{3})$
12. Power	:	HP	= joule per second (or watt)	$\times (7.57 \times 10^{2})$
	:	foot-pound per minute	= joule per second (or watt)	$\times (2.259 \times 10^{-2})$
13. Specific heat	:	BTU/lb °F	= joule per kilogram kelvin (or J/kgK)	$\times (4.184 \times 10^{3})$
14. Thermal conductivity	:	BTU in/hr ft^2 °F	= joule per second metre Kelvin (or J/smK) or watt per metre Kelvin (or W/mK)	$\times (1.442 \times 10^{-1})$

Bibliography

BOOKS

1. Carslaw, H.S. and Jaeger, J.C. (1950), "Conduction of Heat in Solids", Oxford Univ. Press, London.
2. Cook, N.H. and Rabinowicz, E. (1963), "Physical Measurements and Analysis", Addison-Wesley, Reading (Mass.).
3. Bowden, F.P. and Tabor, D. (1964), "Friction and Lubrication in Solids", Clarendon Press, Oxford.
4. Tobias, S.A. (1965), "Machine Tool Vibrations", Blackie, London.
5. Cook, N.H. (1966), "Manufacturing Analysis", Addison-Wesley, Reading (Mass.).
6. Kronenberg, M. (1966), "Machining Science and Applications", Pergamon Press, Oxford.
7. Zorev, N.N. (1966), "Metal Cutting Mechanics", Pergamon Press, Oxford.
8. ASM (1967), "Metals Handbook; vol. 3, Machining", ASM, Metals Park (Ohio).
9. Beckwith, T.G. and Buck, N.L. (1969), "Mechanical Measurements", Oxford & IBH, New Delhi.
10. Armarego, E.J.A. and Brown, R.H. (1969). "The Machining of Metals", Prentice-Hall, Englewood Cliff (New Jersey).
11. Bhattacharyya, A. and Ham, I. (1969), "Design of Cutting Tools", SME, Dearborn (Mich.).
12. Boothroyd, G. (1975), "Fundamentals of Metal Machining and Machine Tools", McGraw-Hill, Kogakusha, Tokyo.
13. Lewis, K.B. and Schleicher, W.F. (1976), "The Grinding Wheel", The Grinding Wheel Institute, Cleveland (Ohio).
14. HMT (1980), "Production Technology", Tata McGraw-Hill, New Delhi.
15. Pandey, P.C. and Singh, C.K. (1980), "Production Engineering Sciences", Standard Publishers, Delhi.
16. SME (1983), "Tool and Manufacturing Engineers Handbook : vol. I, Machining", SME, Dearborn (Mich.).
17. Bhattacharyya, A. (1984), "Metal Cutting—Theory and Practice", Central Book Publishers, Calcutta.
18. Doyle L.E. (ed.) (1984), "Manufacturing Processes and Materials for Engineers", Prentice-Hall, Baglewood Cliffs (N.J.).

19. Shaw M.C. (1984), "Metal Cutting Principles", Clarendon Press, Oxford.
20. Ghosh, A. and Malik, A.K. (1985), "Manufacturing Science", Affiliated East-West Press, New Delhi.
21. Dieter, G.E. (1986), "Mechanical Metallurgy", McGraw-Hill, New York.
22. King. R.S. and Hahn, R.S. (1986) "Handbook of Modern Grinding Technology", Chapman and Hall, New York.
23. DeGramo, E.P., Black, J.T. and Kohser, R. (1988), "Materials and Processes in Manufacturing", Macmillan, New Delhi.
24. Malkin, S. (1989), "Grinding Technology—Theory and Application of Machining with Abrasives", Ellis Horwood, Chichester, England.
25. Oxley, P.L.B. (1989), "The Mechanies of Machining", Ellis Horwood, Chichester, England.
26. Bhattacharyya, A. and Mullick, B.K. (1990), "Technology of Machining Systems", Central Book Publishers, Calcutta.
27. Trent, E.M. and Wright, P.K. (2000), "Metal Cutting", Butterworth-Heinmann, Boston.
28. Juneja, B.L. and Sekhon, G.S. (2005), "Fundamentals of Metal Cutting and Machine Tools", Tata McGraw-Hill, New Delhi.

RESEARCH PAPERS

Basic Mechanics

1. Merchant, M.E. (1944), "Basic Mechanics of the Metal Cutting Process", J. of Appl. Mech., vol. 15, p. A-168.
2. Lee, E.H. and Shaffer, B.W. (1951), "Theory of Plasticity Applied to the Problem of Machining", J. Appl. Mech., vol. 18, p. 405.
3. Russell, J.K. and Brown, R.H. (1966), "The Measurement of Chip-flow Direction", Int. J. Mach. Tool Des. Res., vol., 6, p. 129.
4. Shaw, M.C., Cook, N.H. and Smith, P.A. (1952), "The Mechanics of Three-Dimensional Cutting Operations", Trans. ASME, vol. 74, p. 1055.
5. Pugh, H.D. (1958), "Mechanics of Cutting Process", Proc. Conf. on Tech. of Eng. Manufacture, Inst. of Mech. Engrs., London, p. 237.
6. Shaw, M.C., Cook, N.H. and Finnie, I. (1953), "Shear-Angle Relationship in Metal Cutting" Trans ASME, vol. 75, p. 273.
7. Kobayashi, S. and Thomsen, E.G. (1959), "Some Observations of the Shearing Process in Metal Cutting", Trans ASME, vol. 82, p. 25.
8. Palmer, W.B. and Oxley, P.L.B. (1959), "Mechanics of Orthogonal Cutting", Proc. Inst. of Mech. Engrs., vol. 173, p. 623.
9. Okushima, K. and Hitomi, K. (1961), "An Analysis of the Mechanics of Orthogonal Cutting and its Application to Discontinuous Chip Formation", Trans, ASME, J. Eng. for Ind., vol. 83, p. 545.
10. Oxley, P.L.B. (1961), "A Strain Hardening Solution for the Shear Angle in Orthogonal Cutting", Int. J. Mech. Sci., vol. 3, p. 68.

11. Oxley, P.L.B. , and Hatton, A.P. (1963), "Shear-Angle Solution Based on Experimental Shear Zone and Tool-Chip Interface Stress Distribution", Int. J. Mech. Sci, vol. 5, p. 41.
12. Stabler, G.V. (1951), "The Fundamental Geometry of Cutting Tool", Proc. Inst. Mech. Engrs., vol. 165, p. 14.
13. Brown, R.H. and Armarego, E.J.A (1964), "Oblique Machining with a Single-Cutting Edge", Int. J. Mach., Tool Des. Res., vol. 4, p. 9.
14. Connolly, R. and Rubenstein, C. (1967), "The Mechanics of Continuous Chip Formation in Orthogonal Cutting", Int. J. Mach. Tool Des. Res., vol. 8, p. 159.
15. Spaans, C. (1967), "An Exact Method to Determine the Forces on the Clearance Plane", Annals CIRP, vol. 15, p. 463.
16. Choi, C.Y. (1969), "An Experimental Study of Plastic Deformation due to Oblique Cutting", Adv. Mach. Tool Des. Res., Pergamon Press, Oxford, p. 1081.
17. Morcos, W.A. (1972), "A Solution of the Oblique Continuous Cutting Problem in Conditions of Light Friction at the Chip-Tool Interface", J. Eng. for Ind., vol. 94, p. 1124.
18. Lal, G.K. (1982), "An Experimental Study of Oblique Cutting with Single-edged Tools", Int. J. Mech. Tool Des. Res., vol. 22, p. 282.

Tool Geometry

1. Stabler, G.V. (1951), "The Fundamental Geometry of Cutting Tools", Proc. Inst. Mech. Engrs., vol. 165, p. 14.
2. Armarego, E.J.A. (1965), "The Geometry and Specification of Single-Point Lathe Tools", Int. J. Mach. Tool Des. Res., vol. 4, p. 189.

Force Measurement

1. Loewen, E.G., Marshal, E.R. and Shaw, M.C. (1951), "Electric Strain Gauge Tool Dynamometers", Proc. Amer. Soc. Expt. Stress Analysis, vol. 8, p.1.
2. Cook, N.H., Loewen, E.G. and Shaw, M.C. (1954), "Machine Tool Dynamometers", American Machinist, vol. 98, p. 1171.
3. Cook. N.H. and Loewen, E.G. (1956), "Metal Cutting Measurements and their Interpretation", Proc. Amer. Soc. Expl. Stress Analysis, vol. 13, p. 57.
4. Koenigsberger, F., Marwaha, K. D. and Sabberwal, A.J.P. (1958), "Design and Performance of two Milling Force Dynamometers", Prod. Engr., vol. 37, p. 727.
5. Okushima, K. and Hitomi, K. (1961), "A design of a Three-Component Tool Dynamometer", Memoirs of the Faculty fo Engineering, Kyoto University, Japan, vol. 23, p. 321.
6. Levi, R. (1967), "Drill Press Dynamometer", Int. J. Mach. Tool Des. Res., vol. 7, p. 269.
7. Lal, G.K. (1968), "Forces in Vertical Surface Grinding", Int. J. Mach. Tool Des. Res., vol.8, p. 33
8. Yang, C.T. (1968), "Design of Surface Grinding Dynamometers", Trans. ASME, J. Eng. for Ind., vol. 90, p. 127.
9. Shiozaki, S. and Miyashita, M. (1970), "Design and Dynamics of Dynamometers", Annals. CIRP, vol. 16, p. 653.
10. Ehrenveich, E. and Lenz, E. (1971), "Dynamometer for Drilling Force,", ASME Paper No. 71-Prod.-87.

Tool Temperature

1. Jaeger, J.C. (1942), "Moving Sources of Heat and Temperature at Sliding Contact", Proc. Royal Soc. (NSW), vol. 76, p. 203.
2. Trigger, K.J. and Chao, B.T. (1951), "An Analytical Evaluation of Metal-Cutting Temperatures", Trans. ASME, vol. 73, p. 57.
3. Rapier, A.C. (1954), "A Theoretical Investigation of the Temperature Distribution in the Metal Cutting Process", British J. Appl. Phy., vol. 5, p. 400.
4. Weiner, J.H. (1955), "Shear Plane Temperature Distribution in Orthogonal Cutting", Trans. ASME, vol. 77, p. 1331.
5. Chao, B.T. and Trigger, K.J. (1958), "Temperature Distribution at Tool-Chip and Tool-Work Interfaces in Metal Cutting", Trans. ASME, vol. 80, p. 311.
6. Reichenbach, G.S. (1958), "Experimental Measurement of Metal-Cutting Temperature Distribution", Trans, ASME, vol. 80, p. 525.
7. Trigger, K.J., Cambell, R.K. and Chao, B.T. (1958), "A Tool-Work Thermocouple Compensating Circuit", Trans. ASME, vol. 80, p. 541.
8. Boothroyd, G. (1963), "Temperatures in Orthogonal Metal Cutting", Proc. Inst. Mech. Engrs., vol. 177, p. 789.
9. Stephenson, D.A. (1991), "Assessment of Steady-state Metal-Cutting Temperature Models Based on Simultaneous Infrared and Thermocouple Data", Trans. ASME, J. Eng. for Ind., vol. 113, p. 121.

Tool Wear

1. Taylor, F.W. (1907), "On the Art of Cutting Metals", Trans. ASME, vol. 28, p. 31.
2. Solaja, V (1955), "Tool-life Testing by Facing in a Lathe", Wear, vol. I, p. 512.
3. Trigger, K.J. and Chao. B.T. (1956), "The Mechanism of Crater Wear of Cemented Carbide Tools", Trans. ASME, vol. 78, p. 1119.
4. Shaw, M.C. and Dirke, S.O. (1956), "On the Wear of Cutting Tools", Microtecnic, vol. 10, p. 187.
5. Colding, B.H. (1959), "A Three-Dimensional Tool Life Equation—Machining Economics", Trans. ASME, J. Eng. for Ind., vol. 81, p. 239.
6. Loladze, T.N. (1962), "Adhesion and Diffusion Wear in Metal Cutting", J. Inst. of Engrs. (India), Mech. Eng., vol. 43, p. 108.
7. Taylor, J. (1962), "The Tool Wear-Time Relationship in Metal Cutting", Int. J. Mach. Tool Des. Res., vol. 2, p. 199.
8. Henkin, A. and Datsko, J. (1963), "Influence of Physical Properties on Machinability", Trans. ASME, J. Eng. for Ind., vol. 85, p. 321.
9. Murphy, D.W. (1963). "Machinability of Steels", Proc. Conf. on Int. Res. in Prod. Eng., ASME, New York, p. 177.
10. Takeyama, H. and Murata, R. (1963), "Basic Investigation of Tool Wear", Trans. ASME, J. Eng. for Ind., vol. 85, p. 33.

11. Lorenz, G. (1963-64), "Determination of the Tool Life Exponent from Quick Facing Tests", Annals CIRP, vol. 12, p. 217.
12. Bhattacharyya, A. and Ghosh, A. (1965), "Diffusion Wear of Cutting Tools", Adv. Mach. Tool Des. Res., Pergamon Press, Oxford, p. 225.
13. Cook, N.H. and Nayak, P.N. (1966), "The Thermal Mechanics of Tool Wear", Trans. ASME, J. Eng. for Ind., vol. 85, p. 321.
14. Heginbotham, W.B. and Pandey, P.C. (1966), "Taper Turning Tests Produce Reliable Tool-Wear Equations", Adv. Mach. Tool Des. Res., Pergamon Press, Oxford, p. 515.
15. Venkatesh. V.C. (1966), "Diffusion Wear of High-Speed Steel Tools", Adv. Mach. Tool Des. Res., Pergamon Press, Oxford, p. 401.
16. Rubenstein, C. (1967), "The Role of Surface Energy in Metal Cutting", Proc. of Conf. on Machinability, Iron and Steel Institute, London, p. 49.
17. Rao, I.V. and Lal, G.K. (1977), "Tool Life at High Cutting Speeds", Int. J. Mach. Tool Des. Res., vol. 17, p. 235.

Economics

1. Brewer, R.C. (1958), "On the Economics of the Basic Turning Operation", Trans. ASME, vol. 80, p. 1479.
2. Brown, R.H. (1962), "On the Selection of Economical Machining Rates", Int. J. Prod. Res., vol. 1, p. 1.
3. Shaw, M.C. (1965), "Optimum Selection of Machine Tool Speeds and Feeds", Int. J. Mach. Tool Des. Res. vol. 5, p. 25.
4. Armarego, E.J.A. and Russell, J.K. (1966), "Maximum Profit Rate as a Criterion for the Selection of Machining Conditions", Int. J. Mach. Tool Des. Res., vol. 6, p. 15.

Practical Machining Operations

1. Martellotti, M. (1945), "Analysis of the Milling Process", Trans. ASME, vol. 67, p. 233.
2. Oxford (Jr.), C.J. (1955), "On the Drilling of Metals, I: Basic Mechanics of the Process", Trans, ASME, vol. 77, p. 103.
3. Shaw, M.C. and Oxford (Jr.), C.J. (1957), "On the Drilling of Metals, II: The Torque and Thrust in Drilling", Trans. ASME, vol. 79, p. 191.
4. Koenigsberger, F. and Sabberwal, A.J.P. (1961), "An Investigation into the Cutting Force Pulsations During Milling Operations", Int. J. Mach. Tool Des. Res., vol. 1, p. 15.
5. Bera, S.K. and Bhattacharyya, A. (1966), "Mechanics of Drilling Process", J. Inst. Engrs. (India), Mech. Eng., vol. 46, p. 265.
6. Luk, W.K. (1970), "Mechanics of Turning Operation", Int. J. Mach. Tool Des. Res., vol. 10, p. 351.

Grinding

1. Frisch, J. and Thomsen, E.G. (1951), "Residual Grinding Stresses in Mild Steel", Trans. ASME, vol. 73, p. 337.
2. Littman,W.E. and Wulff, J. (1951), "The Influence of the Grinding Process on the Structure of Hardened Steel", Trans. ASME, vol. 73, p. 21.

3. Backer, W.R., Marshall, E.R. and Shaw, M.C. (1952), "The Size Effect in Metal Cutting", Trans. ASME, vol. 74, p. 61.
4. Outwater, J.O. and Shaw, M.C. (1952), "Surface Temperatures in Grinding", Trans. ASME, vol. 74, p. 73.
5. Letner, H.R. (1955), "Residual Grinding Stresses in Hardened Steel", Trans. ASME, vol. 77, p. 1089.
6. Reinchenbach, G.S., Mayer, J.E., Kalpackcioglu, S. and Shaw. M.C. (1956), "The Role of Chip-Thickness in Grinding", Trans. ASME, vol. 78, p. 847.
7. Backer, W.R. and Merchant, M.E. (1958), "On the Mechanics of Grinding Process", Trans. ASME, vol. 80, p. 847.
8. Krabacker, E.J. (1959), "Factors Influencing Performance of Grinding Wheels", Trans. ASME, J. Eng. for Ind., vol. 81, p. 187.
9. Grisbrook, H., Hollier, R.H. and Varley, P.G. (1962), "Related Patterns of Grinding Forces, Wheel Wear and Surface Finish," Int. J. Prod. Res. vol. 1, p. 57.
10. Hahn, R.S. (1962), "On the Nature of the Grinding Process", Adv. Mach. Tool Des. Res., Pergamon Press, Oxford, p. 129.
11. Yoshikawa, H. and Sata, T. (1963), "A Study on the Wear of Grinding Wheels", Trans. ASME, J. Eng. for Ind., vol. 85, p. 39.
12. Armarego, E.J.A. and Brown, R.H. (1965), "On the Size Effect of Metal Cutting", Int. J. Prod. Res., vol. 1, p. 75.
13. Hiller, M.J. (1966), "On a Three-Dimensional Model of the Surface Grinding Process", Int. J. Mach. Tool Des. Res., vol. 6, p. 109.
14. Farmer, D.A., Brecker, J.N. and Shaw, M.C. (1966-67), "Study of Finish Produced in Surface Grinding, Part I: Experimental", Proc. Mech. E., vol. 182, p. 171.
15. Nakayama, K. and Shaw, M.C. (1966-67), "Study of Finish Produced in Surface Grinding, Part II: Analytical", Proc. I Mech. E., vol. 182, p. 179.
16. Scrutton, R.F., Yousef, N. and Bell, L.R. (1969), "The Scraping Action of Conical Tools", Wear, vol. 14, p. 415.
17. Des Ruisseaux, N.R. and Zerkle, R.D. (1970), "Thermal Analysis of the Grinding Process", Trans. ASME, J. Eng. for Ind., vol. 92, p. 428.
18. Komanduri, R. (1971), "Some Aspects of Machining with Negative Rake Tools Simulating Grinding", Int. J. Mach. Tool Des. Res., vol. 11, p. 233.
19. Malkin, S. and Cook, N.H. (1971), "The Wear of Grinding Wheels, Part I: Attritious Wear", Trans. ASME, J. Eng. for Ind., Vol. 93, p. 1120.
20. Malkin S. and Cook. N.H. (1971), "The Wear of Grinding Wheels, Part II: Fracture Wear", Trans. ASME, J. Eng. for Ind., vol. 93, p. 1129.
21. Rubenstein, C. (1972), "The Mechanics of Grinding", Int. J. Mach. Tool Des. Res., vol. 12, p. 127.
22. Sauer, W.J. (1972), "Thermal Aspects of Surface Grinding", Proc. Int. Grinding Conf., Carnegie Press, Pittsburgh, p. 391.
23. Lindsay, R.P. (1973), "On the Surface Finish-Metal Removal Relationship in Grinding", Trans. ASME, J. Eng. for Ind., vol. 95, p. 815.

24. Rubenstein, C. (1973), "The Factors Influencing the Surface Finish Produced by Grinding", Adv. Mach. Tool Des. Res., Pergamon Press, Oxford, p. 625.
25. Lal, G.K. and Shaw, M.C. (1974), "Experiments with Spherical Tools", Wear, vol. 29, p. 153.
26. Snoeys, R., Peters, J. and Decneut, A. (1974), "The Significance of Chip-Thickness in Grinding", Annals CIRP, vol. 23, p. 227.
27. Lal, G.K. and Shaw, M.C. (1975), "The Role of Grain Tip Radius in Fine Grinding", Trans. ASME, J. Eng. for Ind., vol. 98, p. 119.
28. Malkin, S. (1976), "Selection of Operating Parameters in Surface Grinding of Steels", Trans. ASME, J. Eng. for Ind., vol. 99, p. 56.
29. Malkin, S. (1979), "Negative Rake Cutting to Simulate Chip Formation in Grinding", Annals CIRP, vol. 28, p. 209.
30. Basuray, P.K., Sahay, B. and Lal, G.K. (1980), "A Simple Model for Evaluating Surface Roughness in Fine Grinding," Int. J. Mach. Tool Des. Res., vol. 20, p. 265.
31. Pande, S.J., Halder, S.N. and Lal, G.K. (1980), "Evaluation of Grinding Wheel Performance", Wear, vol. 58, p. 237.
32. Malkin, S. (1984), "Grinding of Metals: Theory and Application", J. Appl. Metal Working, vol. 3, p. 93.
33. Srivastava, A.K., Sriram, K. and Lal, G.K. (1988), "A Simple Analysis for Evaluating Grinding Wheel Loading", Int. J. Mach. Tool Des. Res., vol. 28, p. 181.
34. Srihari, G. and Lal, G.K. (1994), "Mechanics of Vertical Surface Grinding", J. Mat. Proc. Tech., vol. 44, p. 14.

Vibrations

1. Arnold, R.N. (1946), "Mechanism of Tool Vibrations in Cutting of Steel", Proc. Inst. Mech. Engrs., vol. 54, p. 261.
2. Doi, S. and Kalo, S. (1956), "Chatter Vibration of Lathe Tools", Trans. ASME, vol. 78, p. 1127.
3. Tobias, S.A. and Fishwick, W. (1958), "The Chatter of Lathe Tools Under Orthogonal Cutting Conditions", Trans ASME, vol. 80, p. 1079.
4. Thusty, J. and Polack, M. (1963), "The Stability of Machine Tool Against Self-excited Vibrations in Machining", Proc. Int. Res. in Prod. Eng, ASME, p. 465.
5. Albrecht, P. (1965), "The Dynamics of the Metal Cutting Process", Trans ASME, J. Eng. for Ind., vol. 7, p. 152.

Index